PROSPECTS OF HIGHER EDUCATION

EDUCATIONAL FUTURES
RETHINKING THEORY AND PRACTICE
Volume 13

Scope
This series maps the emergent field of educational futures. It will commission books on the futures of education in relation to the question of globalisation and knowledge economy. It seeks authors who can demonstrate their understanding of discourses of the knowledge and learning economies. It aspires to build a consistent approach to educational futures in terms of traditional methods, including scenario planning and foresight, as well as imaginative narratives, and it will examine examples of futures research in education, pedagogical experiments, new utopian thinking, and educational policy futures with a strong accent on actual policies and examples.

Prospects of Higher Education

Globalization, Market Competition, Public Goods and the Future of the University

Edited by

Simon Marginson
Centre for the Study of Higher Education,
University of Melbourne, Australia

SENSE PUBLISHERS
ROTTERDAM / TAIPEI

A C.I.P. record for this book is available from the Library of Congress.

ISBN 978-90-8790-196-7 (paperback)
ISBN 978-90-8790-197-4 (hardback)

Published by: Sense Publishers,
P.O. Box 21858, 3001 AW Rotterdam, The Netherlands
http://www.sensepublishers.com

Printed on acid-free paper

CONTENTS

NOTES ON CONTRIBUTORS

Kate Berniaz served as a project intern in the Centre for Effective Leadership at the American Council on Education in Washington, D.C. in the United States of America. She is currently a graduate student at the University of Victoria studying public policy.
kberniaz@hotmail.com

Elizabeth Bullen is a Research Fellow in the Faculty of Education at Monash University in Melbourne, Australia. She has published on various aspects of gender, education, and popular culture.
elizabeth.bullen@education.monash.edu.au

Gloria Dávila-Casasnovas is the Director of the Policy and Planning Division of the Puerto Rico Council on Higher Education in San Juan, Puerto Rico. Her research focuses on higher education including issues related to gender, economic development, and student access, retention and mobility.
gdavilacasasnovas@gmail.com

Peter D. Eckel is the Director of Programs and Initaitives in the Center for Effective Leadership at the American Council on Education in Washington, D.C. in the United States of America. His scholarship addresses questions of governance and administration, change management, and the external environment.
peter_eckel@ace.nche.edu

Madeleine F. Green is the Vice President, International Initiatives at the American Council on Education in Washington, D.C. in the United States of America. Her work focuses on internationalization of U.S. higher education; advocacy on issues of concern to the international education community in the U.S. and abroad; and cooperation with international partners on higher education issues that have a global impact.
madeleine_green@ace.nche.edu

David M. Hoffman is a doctoral student at the Institute for Educational Research, University of Jyväskylä, Jyväskylä, Finland. His research focus on migration and academic mobility.
david.hoffman@ktl.jyu.fi

Jane Kenway is a Professor of Education in the Faculty of Education at Monash University in Melbourne, Australia. Her research focuses on educational politics and socio-cultural change.
jane.kenway@education.monash.edu.au

Alma Maldonado-Maldonado in an Assistant Professor in the Center for the Study of Higher Education in the College of Education at the University of Arizona, Tucson, United States of America. Her research focuses on international organizations' influence on higher education policy in developing countries, and on university partnerships in the borderlands of Mexico and the U.S.
maldona2@u.arizona.edu

Simon Marginson is a Professor of Higher Education in the Centre for the Study of Higher Education at the University of Melbourne, Australia. His research and scholarship are focused on higher and international education, and education policy, in the context of globalization.
s.marginson@unimelb.edu.au

Tim May is Lead Director of the Centre for Sustainable Urban and Regional Futures, University of Salford, and with colleagues, conducts research into urban futures, science policy and knowledge and regionalization.
t.may@salford.ac.uk

Rajani Naidoo is a Senior Lecturer in the School of Management at the University of Bath and directs the Doctor of Business Administration (Higher Education Management) program. She works on higher education as a global commodity, transnational higher education and higher education in developing countries
r.naidoo@bath.ac.uk

Imanol Ordorika is a Professor of Social Sciences and Education in the Instituto de Investigaciones Económicas at the Universidad Nacional Autonoma de Mexico, Mexico City, Mexico. He works on the political processes and social movements in higher education as well as the impact of globalization on diverse university traditions at the local and international level.
ordorika@servidor.unam.mx

Michael Peters is a Professor of Education in the Department of Educational Policy Studies at the University of Illinois (Urbana-Champaign) and Faculty of Education at the University of Glasgow. He has research interests in educational theory and policy, and in contemporary philosophy.
mpet001@ad.uiuc.edu

Dr. Brian Pusser is an Assistant Professor in the Center for the Study of Higher Education of the Curry School of Education at the University of Virginia. His research focuses on the politics of higher education, the organization and governance of postsecondary institutions, and the role of international, national and state policies in shaping the postsecondary arena.
bp6n@virginia.edu

Gary Rhoades is a Professor of Higher Education and director of the Center for the Study of Higher Education in the College of Education at the University of Arizona, Tucson, United States of America. His research focuses on the restructuring of professions and institutions in academe.
grhoades@email.arizona.edu

Simon Robb is a Research Fellow in the Faculty of Education at Monash University in Melbourne, Australia. He is author of the fictocritical book *The Hulk.*
simon.robb@education.monash.edu.au

Kirsten Turner is an Assistant Dean at the University of Kentucky, Lexington, United States of America. She works on the public benefits derived from greater access to postsecondary education.
jkturn1@email.uky.edu

Jussi Välimaa is a Professor at the Institute for Educational Research, University of Jyväskylä, Jyväskylä, Finland. His research is focused on the social dynamics taking place inside higher education institutions and with society.
jussi.valimaa@ktl.jyu.fi

Martín Velazquez is a doctoral student at the Center for the Study of Higher Education in the College of Education at the University of Arizona, Tucson, United States of America. He is a former campus president of the Leon campus of the Instituto Tecnológico y de Estudios Superiores de Monterrey, in México (ITESM).
mvelazqz@itesm.mx

Waldemiro Vélez Cardona is a Professor of Social Sciences in the Faculty of General Studies in the Río Piedras Campus of the University of Puerto Rico. His research focuses on higher education finance; and the impact of globalization and neo-liberalism in higher education.
waldemirov@hotmail.com

Stéphan Vincent-Lancrin is an Analyst at the Organization for Economic Cooperation and Development (OECD) Centre for Educational Research and Innovation in Paris.
stephan.vincent-lancrin@oecd.org

Leonard J. Waks is a Professor of Educational Leadership and Policy Studies at Temple University in Philadelphia, USA. His research deals with the emerging educational arrangements of global network society.
ljwaks@yahoo.com

SIMON MARGINSON

PREFACE: THE OPEN HORIZON

Higher education and knowledge have never been more generative or more crucial to the human societies in which they are located. Nor have their institutions been more busy, or more troubled. What does the world want from higher education, and what kind of world is it making? Can we talk about a global university market and what does this mean for the social agenda? Where is higher education going?

Accelerated globalization is remaking the prospects and potentials of higher education, especially its cultural potentials, while reinventing government, business and communications everywhere. It has varying meanings. In the USA it underpins the growth of the for-profits, visa issues on the desks of university presidents and debate about higher education as a public or a private good. In Singapore it drives the negotiation of partnerships with major universities from around the world. In Europe it means Bologna and the changes in mobility, program structures and research. It many countries it means the fast growing role of English in teaching and publishing and the roll-out of information and communications technologies.

The global transformation of higher education is now being reinforced by a heightened globalism in the broader political economic environment in the wake of the world-wide focus on climate change. If there are common solutions to be found research and higher education will be central to these. This common crisis is turning the vision of universities outwards, to the state of the world. In the last two decades on campus much energy has been absorbed by the epochal collision between on one hand neo-liberal policy and the new public management (NPM), on the other hand the Deweyan conception of education in democracy, and traditional scholarly cultures. These debates are still being played out. But they are now overshadowed by the many implications of global convergence for local practices, national systems and ways of thought. These include the transfer of neo-liberalism and the NPM and much more across national borders. Global university rankings are a potent (and uncomfortable) reminder that in higher education, which always was partly internationalised, the ultimate horizon is now the global one.

The NPM and global convergence overlap but these are distinct problematics. The discussion about the NPM is about what is being done *to* higher education by Treasury departments and executive managers, from outside the sector as it were. Much of the discussion about globalization has been couched in similar terms, but this is misplaced. Higher education institutions, especially research universities, are not just objects of globalization but among its primary subjects. If globalization is about cross-border flows and systems of people, messages, ideas, knowledge, technologies and capital, then higher education institutions and people who work in them are instigating many of the flows themselves. If globalization is also about world-wide inequalities and injustices - and in higher education and other sectors it is - then universities are both part of the problem, and part of the solution. What this means for teaching and learning, and research and scholarship, and policy and organization in higher education, are among the preoccupations of this book.

Global convergence means that processes conducted in isolation in localities and nations no longer absorb us as they once did. It piles uncertainty upon uncertainty because there is a myriad of possible futures. It brings a new scope for agency and contingency and exciting forms of freedom. It also heightens the capacity of strong universities and nations to dominate the rest. The barriers are being stripped away. Direct cross-border relations subvert the old distinctions between national and international, and remake the relations between government and higher education institutions. National funding and regulation have not and will not disappear and remain determining, but national policy is relativised in the global setting. If the global dimension is now the horizon of possibility, it is an open horizon.

CONTENTS OF THIS BOOK

The changing prospects and potentials of knowledge formation and learning, taking in universities and other higher education institutions, are a complex intellectual problem. But it *can* be understood. This problem is no more difficult than mapping the human genome or modelling the global climate (though it must be said that less research resources are put into solving it). We have yet to fashion all of devices that we need to make such an analysis. The central purpose of this book, which draws on leading scholars in higher education studies from around the world, is to develop imaginative frameworks and conceptual tools for this project.

However, in grappling with the impact of the global we are not beginning from scratch. We have inherited certain concepts from political philosophy, sociology and geo-spatial studies that are useful here, though they might need to be remade. This book is premised on two principal distinctions, which are explained at more length in chapter 2. The first is the distinction between *public goods and private goods*. Much of the scholarly discussion has been about 'states' and 'markets' in higher education, and it is often couched in terms of the classical liberal opposition between the two. It is true that both states and markets are heavily implicated in the many changes now occurring, but they have limits as central explanatory devices. First, they imply closed systems bound by an internal logic. Modelling states and markets (especially markets) this way has much to say about the tendencies they create but has drawbacks when the models are used to frame observation. In reality matters are more fragmented and open, states and markets do not between them cover the whole of human endeavour, and the closure makes it difficult to map relations between the two which are inter-penetrated at many points. Second, states and markets are essentially mediums and means of human action, not ends in themselves. What is more important is the social effects produced within, from or by, states and economic markets and positional competition – and also in other realms which do not necessarily fit into state or market, including private life, the the realm of the imagination, and the global space. Arguably, the notions of public goods and private goods, which cut across the state/market distinction, cover more ground. At the same time the public goods/private goods coupling is useful for understanding, measuring and judging the work of states, markets and higher education institutions. Some chapters in this book focus primarily on private goods and market competition, some mostly focus on public goods, some address both.

The second distinction is between the *global, national and local dimensions* or spheres of action. 'Global' refers to spaces, systems, relations, elements, agents and identities constituting and constituted by the world as a whole. The 'global' dimension does not include all national and local elements, only those contributing to the world as an integrated world (we can use 'world-wide' for the universal notion). Amid the first open information environment in history, national policy and local culture can no longer be isolated. But the global dimension is not all-consuming, nor are global effects the same everywhere. Many institutions have no obvious global connections. The quantity and quality of engagement varies from nation to nation. Some are more insulated than others. Global flows between different nations, and between different local institutions across borders, can be two-way or uni-directional. The world as envisioned from hegemonic USA is not the world envisioned from the marginal zones. Differing visions, and differing patterns of movement, emerge in the chapters of the book.

However, the main contribution that *Prospects of higher education* makes to the field is to bring together the analytical frame of public/private goods with the analytical frame of global/national/local, allowing us to explore public and private (and markets and states) in more than one dimension at the same time. This is where the world has gone and it is time scholars caught up. This move allows us to do new things in the study of higher education. For example, it allows us identify global public goods in education and research, which have not been discussed; to map the wash-back effects of global trade in private educational goods on the stratification of national systems; and rethink the local social role of institutions. The three dimensions also help us in developing different geo-spatial perspectives on higher education, different vantage points for observation and analysis.

Part I consists of chapters conceived from a world-wide and global perspective. Using coordinates such as global/national, state/non state provision, self-centred/socially responsive and high/low participation Stéphan Vincent-Lancrin (chapter 1) imagines possible futures for higher education. Simon Marginson (chapters 2 & 3) maps the world-wide landscape, including the formation of global markets and global public goods, while keeping national systems in mind, and the rise of global university rankings and the response. Leonard J. Waks (chapter 4) brings the large industry of corporate universities, partly engendered by globalization, out of the shadows. Jane Kenway, Elizabeth Bullen and Simon Robb (chapter 5) explore the implications of knowledge economy policies for academic freedom and imagining.

Part II consists of chapters from and about the cross-border relations in the Americas, which occupy a special position in world-wide higher education because of the global dominance of the USA, its hispanization, and the role of the nations on the geo-strategic periphery of the USA in that. Peter Eckel, Madeleine Green and Kate Berniaz (chapter 6), writing from the American Council on Education, present a framework for charting the programs provided outside the United States by U.S. institutions to non citizens. Brian Pusser, Waldemiro Vélez-Cardona and Gloria Dávila-Casasnovas (chapter 7) describe what happened when the Pell-grant financed U.S. student market was transplanted raw into the post-secondary system of Puerto Rico. Gary Rhoades, Alma Maldonado-Maldonado, Imanol Ordorika and Martín Velazquez (chapter 8), spanning the USA and Mexico, re-imagine public and private universities in terms of social mission rather than academic capitalism.

Part III shifts the perspective to the two other dimensions: national systems, and the regions and localities within them. In chapter 9 Jussi Välimaa and David M. Hoffman provide a historically grounded account of universities in Finland where the national policy horizon has been displaced by the global horizon. Michael Peters and Tim May (chapter 10) discuss the regional knowledge economy that has emerged in the North-West of the UK in response to national policies and global developments. Rajani Naidoo (chapter 11) uses Bourdieu's theorisation of the field of higher education, drawn on also in chapter 2, to explore social competition and system stratification in post-apartheid South Africa. Kirsten Turner and Brian Pusser (Chapter 12) draw out the dilemmas of the elite public sector (state) university in the USA, the principal provider of social leaders and positional goods, while at the same time being responsible for the democratic provision of equitable access for all citizens.

THANKS AND ACKNOWLEDGEMENTS

I am grateful to the Australian Research Council for the freedom afforded by a Professorial Fellowship, which brought with it the time and the welcome obligation to push deeper into the question of the global. It has been a pleasure to work on this edited collection because of the collaborative relationships that underlie the various contributions. It was a good experience to prepare a book that includes close and long-standing friends, including part of the 'whiskey seminar' that gathers every year in C class conference hotels in mid-priced towns in the United States for the meeting of the Association for Studies in Higher Education, and reflects in formal and informal session on American hegemony in the sector, the differentiation of commodities in the hospitality industry, and the apalling design of the carpet in the foyer. My sincere thanks to all the contributors, who responded quickly and well to my requests, to Michael Peters who created this project, and to Peter de Liefde who has been an exemplary publisher.

I would also like to thank Richard James and my other colleagues at the Centre for the Study of Higher Education at the University of Melbourne, who make it such a fine place to work, and to Glyn Davis for his enthusiasm about reflective scholarship on higher education. My appreciation is also for Mum and Dad and

Melba for the kindness of recent years, and of course for our beloved daughter. In the preface to *Educating Australia* (1997) I remarked that Ana Rosa had joined us towards the end of the writing and that she 'is even more wonderful than writing books'. Now Ana Rosa loves books herself. There is a lot of happiness in all that.

Melbourne, 20 June 2007

Chapter 11 by Rajani Naidoo was originally published as 'Fields and institutional strategy: Bourdieu on the relationship between higher education, inequality and society, in the *British Journal of Sociology of Education*, 25(4), 2004, pp. 446-472. It is reprinted with the kind permission of the publisher, Taylor and Francis. Chapters 1, 4, 5, 8 and 10 were originally published the *Journal of Policy Futures in Education*, 2(2), 2004. These articles are reprinted with the kind permission of the authors, with slight changes for editing and updating. The original articles were: Stephan Vincent-Lancrin, 'Building future scenarios for universities and higher education: An international approach' (245-262); Michael A, Peters and Tim May, 'Universities, regional policy and the knowledge economy' (263-277); Leonard J. Waks, 'In the shadow of the ruins: Globalization and the rise of corporate universities' (278-298); Gary Rhoades, Alma Maldonado-Maldonado, Imanol Ordorika & MartínVelazquez, 'Imagining alternativas to global, corporate, new economy academic capitalism' (316-329); Jane Kenway, Elizabeth Bullen & Simon Robb, 'The knowledge economy, the techno-preneur and the problematic future of the university' (330-349).

PART I

PERSPECTIVES OF THE WORLD

'We were flying over America and suddenly I saw snow, the first snow we ever saw from orbit... I had never visited America, but I imagined that the arrival of autumn and winter is the same there as in other places, and the process of getting ready for them is the same. And then it struck me that we are all children of our Earth'.

- attributed to Soviet Russian cosmonaut Aleksandr Pavlovich Aleksandrov, Soyuz T-9 mission, 1983

STÉPHAN VINCENT-LANCRIN[1]

1. BUILDING FUTURE SCENARIOS FOR UNIVERSITIES AND HIGHER EDUCATION

An International Approach

INTRODUCTION

Although the university as we know it has undergone many evolutions over its long existence, today many forces are putting it under pressure to change. It has entered a new period of questions and doubts. While this is certainly not the first time in its history that this has occurred (for example Newman, 1854; Clarke, 1963), it underlines the present challenges faced by liberal education and universities in general (Maskell & Robinson, 2001; Graham, 2002; Scott, 2002). In many OECD countries, the university has already changed significantly over the past decades, but these changes should become greater in the future because of the rapid evolution of the post-secondary education landscape. It is timely, even urgent, to consider the possible future of the university, to create a common understanding of the socio-economic changes affecting it and to help post-secondary education stakeholders propose adequate responses to these changes. The urge to reflect anew on the future of the university and of higher education is also highlighted by a number of ongoing projects on the future of higher education (for example de Boer et al, 2002; Newman, 2000), by a increasing amount of literature on this subject (for example Dudderstadt, 2000; Renaut, 2002; Altbach, 1998), as well as by new policy papers in several OECD countries reflecting on future directions for higher education (for example DfES, 2003; European Commission, 2003; HEA, 2003).

Basically, four features of the current changes must be stressed because of the strong impact they have on the university. First, post-secondary education systems have become increasingly large and encompass a variety of needs that higher education systems confined to an 'elite' audience did not have to deal with in the past. The missions and the position of the university in post-secondary systems and, more widely, in society, have thus evolved and need to be reconsidered. Second, deliberately or not, greater autonomy for universities and the entry of new private providers often accompanied this evolution. Third, although post-secondary education systems have always been very different across the world, the different

---

[1] The insights and encouragement provided by CERI colleagues Kurt Larsen, Keiko Momii, Riel Miller, and David Istance are gratefully acknowledged, as well as the helpful comments and debates at the OECD/Japan forum on 'The Future of Universities' held in Tokyo in December 2004: http://www.simul-conf.com/oecd_japan/index_en.html. The author is involved in an ongoing CERI project on the future of tertiary education and has benefited from useful feedback on various occasions. The views expressed are the author's and are not necessarily those of the OECD and its member countries. Contact: Stephan.Vincent-Lancrin@oecd.org.

Simon Marginson (ed.), Prospects of Higher Education, 3–27.

directions they take may raise new issues in a more globalized and liberalised world. Four, information and communication technologies have started to open new avenues for teaching and research in higher education. All these changes invite us to examine with a fresh eye the missions and role of the university in our changing world, in order to help avoid undesirable situations and to encourage the desirable moves of post-secondary systems. This is particularly important at a time when higher education receives much attention, with numerous reforms in higher education being considered in the OECD area.

The paper presents a set of scenarios for universities and higher education in the OECD area. The first section gives a brief overview of the main forces currently at play in higher education in OECD countries, setting the context in which speculation about the future takes place. The second section discusses the purposes of a scenario-building exercise, the methodology used and (some of) its caveats. The third section presents the set of scenarios, before some concluding remarks.

DRIVING FORCES

What are the driving forces that are currently challenging and that may, in the future, remodel higher education systems in OECD countries? They can be found in at least five interrelated fields: the demography of and participation in education; internationalisation; technology; the knowledge economy; and governance.

Demographic and Participation Trends

In most OECD countries, there has been a rising demand for post-secondary education over the past decades and a significant increase of the tertiary enrolment rates, resulting in 'mass education'. However, the picture might change in the decades to come. Unless there is an increase in tertiary education participation rates, the decline in the number of young people in most OECD countries should lead to a decrease in the number of young domestic students (if current trends continue). Table 1.1 gives an overview of the expected demographic changes in the school age population. A continuous decline in the number of young people is forecast. The younger the age, the higher the predicted demographic decline. The traditional university-age population (19-25) may thus decrease continuously until about 2030 (when the 2015 group of 5-year-olds will enter university).

On the other hand, an increasing demand for post-secondary education coming from other types of students may compensate for the demographic decline of the traditional university-age population: older students already engaged in active life, retired people and students coming from the transition and developing worlds.

Participation in tertiary education is considered as 'universal' when participation rates reach about 50 per cent. While the length of tertiary studies can still increase, participation is already high in the OECD area, for both males and females. Levels of participation in tertiary education are uneven among OECD countries, but every second young person (53 per cent) in the OECD area will enter general higher education programmes during his/her lifetime, assuming that current entry rates

Table 1.1. Relative size and expected changes in school-age populations (2000)

Size of school-age populations as a percentage of total population and school-age population projections

	Percentage of the population			*Index change in school-age population (base year 2000 = 100)*					
				Ages 5-14		*Ages 15-19*		*Ages 20-29*	
	Ages 5-14	*Ages 15-19*	*Ages 20-29*	*1990*	*2015**	*1990*	*2015**	*1990*	*2015**
	1	2	3	4	5	6	7	8	9
OECD countries									
Australia	14	7	15	92	99	102	102	97	106
Austria	12	6	13	96	68	108	86	131	93
Belgium	12	6	13	99	79	108	92	118	96
Canada	13	7	14	93	89	94	96	113	109
Czech Republic	12	7	17	126	71	121	66	80	73
Denmark	12	5	14	91	88	134	122	113	93
Finland	12	6	12	101	76	91	88	115	103
France	13	7	14	103	97	108	94	107	93
Germany	11	6	12	94	75	97	89	140	101
Greece	10	7	15	128	84	110	74	95	71
Hungary	12	7	16	124	70	123	76	81	74
Iceland	16	8	15	95	86	100	99	102	105
Ireland	15	9	16	122	114	98	80	79	89
Italy	10	5	14	112	84	141	87	114	70
Japan	10	6	15	129	92	133	82	91	69
Korea	14	8	17	118	89	117	81	103	79
Luxembourg	12	5	13	80	109	88	124	107	130
Mexico	23	11	19	98	97	100	105	77	103
Netherlands	12	6	13	91	85	118	106	124	99
New Zealand	15	7	14	87	91	107	109	108	113
Norway	13	6	14	88	85	120	113	111	103
Poland	14	9	15	123	68	85	59	86	89
Portugal	11	7	16	129	96	128	86	92	69
Slovakia	m	m	m	120	71	96	65	85	85
Spain	10	7	16	138	84	131	71	98	62
Sweden	13	6	13	84	61	112	89	111	108
Switzerland	12	6	12	91	70	107	94	139	102
Turkey	20	10	19	106	99	91	103	77	99
United Kingdom	13	6	13	93	80	104	98	119	105
United States	14	7	13	87	96	89	109	111	120
OECD mean	*13*	*7*	*15*	*105*	*85*	*109*	*91*	*104*	*94*
WEI participants									
Argentina	18	9	16	98	106	85	106	78	110
Brazil	22	11	18	104	98	86	91	88	104

Chile	19	8	16	87	97	96	112	103	118
China	18	7	16	91	82	122	95	107	99
Egypt	23	12	17	88	96	71	105	82	138
India	25	9	17	88	101	85	112	85	125
Indonesia	20	11	19	99	94	91	98	83	107
Jamaica	20	10	18	101	92	93	93	94	111
Jordan	27	12	20	78	143	72	139	57	124
Malaysia	22	10	18	86	102	79	116	87	132
Paraguay	26	11	16	78	122	71	127	81	154
Peru	23	10	18	95	100	87	105	81	117
Philippines	24	11	18	85	105	82	118	80	132
Russian Federation	13	8	15	114	62	86	53	101	93
Thailand	16	9	19	108	99	102	99	94	92
Tunisia	22	11	18	102	92	83	79	83	108
Uruguay	16	8	16	99	102	102	109	87	102
Zimbabwe	25	12	20	78	116	74	129	79	156
WEI mean	*21*	*10*	*17*	*93*	*100*	*87*	*105*	*86*	*118*

* These figures are projections.
Sources: OECD/UIS WEI and UN Population Division.

continue. On average in OECD countries, a 17-year old can expect to receive 3.0 years of tertiary education. The percentage of the population of 25 to 34-years-old attaining tertiary education increased from 21 to 30 per cent between 1991 and 2001 (for the 19 OECD countries for which data were available for both years) and amounted to 31 per cent on average in the OECD area in 2004. The average age of university students has increased, as a result of greater participation by older students in comparison with high-school leavers, but university students under 30 still represent more than 80 per cent of enrolments in most OECD countries. In many countries 80 per cent of the students enrolled in general tertiary education are under 23. Maintaining current capacity in tertiary education may require a change in the age structure of the student population (Table 1.2).

While the traditional younger demand for tertiary education may decrease in the OECD area, it may increase significantly in transition and developing countries. Indeed, it is forecast that, in several OECD countries, the school-age population will continue growing until 2030 (Table 1). Although considerable progress has been achieved in participation in and access to education in developing countries, it remains limited, especially at the upper secondary and tertiary levels ultimately crucial for economic development (World Bank, 2002). While there is already unmet demand in transition countries, participation in tertiary education is likely to continue to expand over the next decades in developing countries, as the objectives of universal primary education is achieved. As for university expectancy, only one in four young persons (27 per cent) was expected to enter general higher education programmes during his/her lifetime in 2000, assuming that current entry rates continue. In 2000, on average a 17-year old from a country participating in the

World Education Indicators programme[2] could expect to receive 1.2 years of tertiary education — less than half as much as in the OECD area that year.

In order to deal with a large unmet demand, many transition and developing countries increasingly use international student mobility as well as international educational programmes and institutions to build their capacity in higher education. International student mobility to the OECD area doubled between 1980 and 2001 and the number of foreign students enrolled in the OECD area continues to rise quicker than total tertiary enrolments. Enrolments of international students through foreign branch campuses, joint ventures with local institutions or franchise arrangements are also likely to expand significantly. Information and communication technologies, as well as the liberalisation of trade and investment, facilitate this trend towards globalization of demand and supply in post-secondary education (OECD, 2004). International expansion might assist universities from the OECD area to face the expected demographic decline in their countries.

How will pedagogy, organization of curricula and provision be affected by the changes in number, age and origin of the student population? How will the traditional missions of universities evolve?

Governance and Funding

Shifts in funding and governance could also affect the missions and organization of the university sector. Higher education systems vary considerably within the OECD area in terms of funding and governance and display differing mixes of private and public provision. However, in most OECD countries the typical higher education institution is public or private not-for-profit, and the higher education sector is largely publicly funded. In nine of the 28 OECD countries from which data are available, private income exceeds 30 per cent of the total income of tertiary educational institutions: Korea (77 per cent), the United States (67 per cent), Japan (60 per cent), Australia (52 per cent), Canada (44 per cent), New Zealand (38 per cent), Poland (31 per cent), Mexico (31 per cent) and the United Kingdom (30 per cent). In the context of an ageing society, possibly leading to a reinforcement of health issues and pensions as top public priorities, universities might, in the future, have reasons to raise a greater share of their funds from a variety of private sources, including higher tuition fees. While the GDP per capita has increased since 1995 in all OECD countries, public expenditures per student on tertiary educational institutions have decreased in seven of the OECD countries for which information is available (for both years), that is Poland (-37 per cent), Australia (-30 per cent), the Slovak Republic (-25 per cent), the Czech Republic (-14 per cent), Sweden (-6 per cent), the United Kingdom (-12 per cent) and Portugal (-5 per cent). However, in the UK and Sweden the drop in public funding has been fully compensated by an increase in private funding (Table 1.3).

---

2 Nineteen countries participate in the OECD/UNESCO World Education indicators (WEI) programme: Argentina, Brazil, Chile, China, Egypt, India, Indonesia, Israel, Jamaica, Jordan, Malaysia, Paraguay, Peru, Philippines, Russian Federation, Thailand, Tunisia, Uruguay and Zimbabwe.

Research and development have also been increasingly funded by private sources over the two last decades, but not research carried out by the higher education sector. Gross domestic expenditure on research and development amounted on average to 2.1 per cent of GDP in the OECD in 2001. While those expenditures increased by 0.1 per cent of GDP on average between 1981 and 2000, the share of private expenditures increased by 0.3 per cent to 1.3 per cent of GDP and the share of government expenditures decreased by 0.2 per cent to 0.6 per cent of GDP.[3]

However, while private funding of research has overall remained stable or slightly increased in several OECD countries, it did not amount to more than 10 per cent of the university expenditures on research except in three OECD countries (for which information is available) in 2001: Korea (14.3 per cent excluding R&D in the social sciences and humanities), Germany (11.3 per cent) and Iceland (10.9 per cent). The average share of the business enterprise and higher education sectors in total performed research in the OECD area rose between 1981 and 2000, from 66 per cent to 69 per cent, and 16 per cent to 17 per cent, respectively. While the private not-for-profit sector carried out a stable share of total research over the period (about 3 per cent), the share of total research performed by the government sector (non higher education: military, agriculture and so on) decreased by 5 per cent to 10 per cent. Not surprisingly, basic research represented an average of 59 per cent of the total research performed by the higher education sector in the OECD area in 2001. This highlights two trends over the period: first, the drop in research undertaken by the government sector and of government funding for research in the higher education sector; second, the increase in research carried out and financed by the private sector. But university research remains largely government-funded (Table 1.4).[4]

In a context of mass tertiary education and ageing society, the best way to fund and deliver both research and teaching components of higher education is under debate in many OECD countries. Concerning levels as well as sources of funding, the debates include consideration of, among other factors, national budget priorities and the desire to increase the available resources; questions about the efficiency of resource use; ensuring that public policy objectives (for example high-quality education and research) are met; and determining what government should provide and how costs should be shared among different groups in the society (taxpayers, students and families, companies). Moreover, there is a strong social demand for better public management. Accountability, transparency, efficiency and effectiveness, responsiveness and forward vision are now considered as the principal components of good public governance, which universities are and will increasingly be asked to implement (Braun & Merrien, 2002). The shift towards more autonomy and entrepreneurship is a common trend in higher education management in most OECD countries (OECD, 2003; Marginson & Considine, 2000; Etzkowitz et al., 2000; Martin, 2002).

---

3 For OECD countries for which data are available for both years: total expenditures rose from 1.87 to 1.98 per cent of GDP; private expenditures from 1.00 to 1.31 per cent of GDP; government expenditures fell from 0.87 to 0.66 per cent.

4 Source for all data in this section: OECD R&D Database.

Table 1.2. Entry rates and expected years of schooling at university post-secondary level, 2004

	2004		*Change in expectancy all levels combined*	*2004*								
	Education expectancy - Full-time and part-time		*1995-2004 (1995=100)*	*Tertiary-type B (Vocational)*			*Tertiary-type A (General)*					
	All levels education combined	*Tertiary education*		*Net entry rates Sum of net entry rates for each year of age*			*Net entry rates Sum of net entry rates for each year of age*			*Age at:*		
	M+F	M+F	M+F	M+F	M	F	M+F	M	F	20th percentile[1]	50th percentile[1]	80th percentile[1]
	(1)	(3)	(4)	(1)	(2)	(3)	(4)	(5)	(6)	(7)	(8)	(9)
OECD countries												
Australia*	20.7	3.6	107	m	m	m	70	65	74	18.6	20.9	27.4
Austria*	16.3	2.3	101	9	8	10	37	33	41	19.3	20.6	23.3
Belgium*	19.6	3.0	109	35	28	42	34	33	35	18.3	18.9	22.4
Canada	m	2.9	m	m	m	m	m	m	m	m	m	m
Czech Republic	17.0	2.1	119	10	7	13	38	36	41	19.5	20.4	22.6
Denmark	19.0	3.2	112	21	20	21	55	43	68	20.9	22.6	27.1
Finland*	20.0	4.5	116	a	a	a	73	65	82	19.8	21.5	27.3
France	16.8	2.8	102	m	m	m	m	m	m	m	m	m
Germany*	17.4	2.3	106	16	13	19	37	38	37	20.1	21.4	24.1
Greece	16.9	3.9	121	26	26	27	33	30	37	18.1	18.6	19.3
Hungary*	17.6	2.9	122	9	7	11	68	61	76	19.2	20.9	27.6

Iceland	19.7	3.5	118	8	8	8	79	56	102	20.9	23.5	<40
Ireland*	17.2	2.9	112	17	18	15	44	39	50	18.3	19.1	20.0
Italy*	17.0	2.9	m	1	1	1	55	49	62	19.2	19.8	22.1
Japan	m	m	m	32	24	41	43	49	36	m	m	m
Korea	16.6	4.3	113	46	44	48	48	52	45	m	m	m
Luxembourg	14.4	m	m	m	m	m	m	m	m	m	m	m
Mexico	13.4	1.2	111	2	2	1	29	28	29	18.4	19.6	23.7
Netherlands	17.4	2.7	m	a	a	a	56	52	61	18.4	19.8	22.7
New Zealand	19.1	4.2	m	51	45	57	89	74	104	18.9	21.9	<40
Norway	18.4	3.6	105	1	1	1	69	58	80	20.0	21.2	29.0
Poland	17.0	3.3	118	1	n	1	71	66	76	19.5	20.4	22.9
Portugal	17.1	2.6	103	m	m	m	m	m	m	m	m	m
Slovak Republic	15.7	1.9	m	2	1	4	47	42	52	19.5	20.9	26.9
Spain*	17.2	3.0	101	22	20	23	44	37	52	18.4	19.1	22.4
Sweden	20.3	3.8	113	8	8	8	79	64	94	20.3	22.8	<40
Switzerland	16.8	2.1	m	17	20	14	38	39	38	20.0	21.6	26.1
Turkey*	12.6	1.5	133	16	19	13	26	29	22	18.6	20.0	23.6
United Kingdom	20.7	2.8	121	28	m	m	52	m	m	18.8	22.4	25.6
United States	16.0	4.1	m	x(4)	x(5)	x(6)	63	56	71	19.4	21.2	24.0
Country mean	*17.4*	*3.0*	*113*	*16*	*14*	*16*	*53*	*48*	*59*	-	-	-
Non-OECD												
Brazil 1	16.7	1.3	m	2	3	2	47	42	53	19.7	23.7	<40
Chile 1	15.0	m	m	25	28	21	46	44	47	m	m	m
Israel	15.6	2.9	m	m	m	m	58	52	64	21.4	23.7	27.8
Russian Fed. 2	15.0	3.6	m	33	x(1)	x(1)	67	x(4)	x(4)	m	m	m

Source: OECD. Note: See OECD, 2006, www.oecd.org/edu/eag2006 for notes

Table 1.3. Distribution of public and private sources of funds for tertiary educational institutions, OECD countries; annual expenditure on educational institutions per student in equivalent US dollars converted using PPPs, based on full-time equivalents; and change in public and total expenditures on educational institutions (1995, 2003)

	2003				1995				Change in funding per student % change 1995-2003	
	Public sources %	Private sources[1] %	Private: of which: subsidised %	Spending per student %	Public sources %	Private sources[1] %	Private: of which: subsidised %	Spending per student %	Public funding %	All sources %
	(1)	(2)	(3)	(4)	(5)	(6)	(7)	(8)	(9)	(10)
Australia	48.0	52.0	0.9	12406	64.6	35.4	8.3	13 084	-30%	-6%
Austria	92.7	7.3	1.6	12344	96.5	3.5	4.6	11 663	10%	15%
Belgium	86.7	13.3	4.7	11824	m	m	m	m	m	m
Canada 2	56.4	43.6	0.9	19992	56.6	43.4	22.3	12 153	28%	28%
Czech Republic	83.3	16.7	m	6774	71.0	29.0	8.6	8 031	-14%	-26%
Denmark3	96.7	3.3	m	14014	99.4	0.6	n	11 607	14%	18%
Finland	96.4	3.6	n	12047	m	m	m	8 633	6%	7%
France	81.3	18.7	2.3	10704	84.3	15.7	2.6	7 411	m	m
Germany	87.1	12.9	n	11594	92.8	7.2	a	9 765	6%	8%
Greece	97.4	2.6	m	4924	m	m	m	3 301	26%	26%
Hungary	78.5	21.5	n	8576	80.3	19.7	n	7 656	5%	7%
Iceland	88.7	11.3	n	8023	93.0	7.0	m	m	m	m
Ireland	83.8	16.2	4.2	9341	69.7	30.3	m	7 189	49%	21%
Italy	72.1	27.9	4.5	8764	m	m	m	6 422	10%	28%

Japan3	39.7	60.3	m	11556	42.0	58.0	m	9 349	7%	14%
Korea	23.2	76.8	0.7	7089	m	m	m	m	m	m
Luxembourg	m	m	m	m	m	m	m	m	m	m
Mexico	69.1	30.9	0.8	5774	77.4	22.6	m	4 713	1%	13%
Netherlands	78.6	21.4	1.5	13444	79.9	20.1	m	11 520	0%	3%
New Zealand	61.5	38.5	m	8832	m	m	m	m	m	m
Norway	96.7	3.3	m	13772	93.6	6.4	n	14 088	m	0%
Poland4	69.0	31.0	m	4589	m	m	m	m	-37%	-37%
Portugal	91.5	8.5	m	7200	96.5	3.5	m	4 415	-5%	-5%
Slovak Republic	86.2	13.8	m	4678	85.3	14.7	m	m	-25%	-17
Spain	76.9	23.1	2.0	8943	74.4	25.6	2.0	4 792	47%	43%
Sweden	89.0	11.0	a	16073	93.6	6.4	a	14 970	-6%	0%
Switzerland	m	m	m	25900	m	m	m	15 671	46%	46%
Turkey	95.2	4.8	m	m	97.0	3.0	0.7	3 375	94%	94%
United Kingdom	70.2	29.8	0.6	11866	63.9	36.1	16.0	10 572	-12%	0%
United States 2	42.8	57.2	m	24074	m	m	m	m	38%	10%
Country mean	*76.4*	*23.6*	*1.5*	*11254*	~	~	*m*	*m*	*m*	*6%*

Notes: 1. Including subsidies attributable to payments to educational institutions received from public sources.
2. To calculate private funds net of subsicies, subtract public subsides (columns 3, 6) from private funds (columns 2, 5).
PPP = Purchasing Power Parity.
Source: OECD, 2006

Table 1.4. Percentage of Gross Domestic Expenditure on Research and Development (GERD) performed by sector, 1981, 2000

		Business Enterprise	*Government*	*Higher Education*	*Private Non-Profit*	*total*
Australia	1981	25.02	45.11	28.55	1.32	100
	2000	47.07	23.10	27.07	2.76	100
Austria	1981	55.85	9.03	32.80	2.33	100
	2000	m	m	m	m	m
Canada	1981	48.11	24.42	26.66	0.82	100
	2000	56.78	11.28	30.98	0.96	100
Czech Rep.	1981	m	m	m	m	m
	2000	59.96	25.32	14.21	0.51	100
Denmark	1981	49.7	22.67	26.74	0.88	100
	2000	m	m	m	m	m
European U.	1981	62.03	18.80	17.81	1.36	100
	2000	64.18	13.59	21.35	0.88	100
Finland	1981	54.66	22.55	22.24	0.56	100
	2000	70.91	10.58	17.85	0.67	100
France	1981	58.92	23.59	16.42	1.07	100
	2000	62.51	17.32	18.75	1.42	100
Germany	1981	68.97	13.44	17.06	0.53	100
	2000	70.33	13.58	16.09	x	x
Greece	1981	22.46	63.08	14.46	x	x
	2000	m	m	m	m	m
Hungary	1981	m	m	m	m	
	2000	44.32	26.09	24.02	5.57	100
Iceland	1981	9.61	60.74	25.97	3.68	100
	2000	56.36	25.48	16.24	1.92	100
Ireland	1981	43.58	39.31	16.03	1.08	100
	2000	m	m	m	m	m
Italy	1981	56.37	25.72	17.91	x	x
	2000	50.07	18.91	31.02	x	x
Japan	1981	65.96	12.02	17.56	4.46	100
	2000	m	m	m	m	m
Korea	1981	m	m	m	m	m
	2000	74.05	13.31	11.28	1.36	100
Netherlands	1981	53.26	20.77	23.18	2.78	100
	2000	57.6	12.8	28.78	0.82	100
Norway	1981	52.87	17.65	28.95	0.52	100
	2000	m	m	m	m	m
Poland	1981	m	m	m	m	m

	2000	36.09	32.25	31.53	0.13	100
Russian Fed.	1981	m	m	m	m	m
	2000	70.78	24.44	4.55	0.22	100
Slovak Rep.	1981	m	m	m	m	m
	2000	65.81	24.68	9.51	0	100
Slovenia	1981	m	m	m	m	m
	2000	56.32	25.90	16.61	1.17	100
Spain	1981	45.49	31.57	22.95	x	x
	2000	53.66	15.82	29.62	0.90	100
Sweden	1981	63.65	6.09	29.99	0.26	100
	2000	m	m	m	m	m
Switzerland	1981	74.20	5.92	19.88	x	x
	2000	73.91	1.31	22.86	1.92	100
Turkey	1981	m	m	m	m	m
	2000	33.44	6.19	60.37	x	x
UK	1981	62.96	20.64	13.55	2.85	100
	2000	65.60	12.17	20.78	1.45	100
USA	1981	71.24	12.46	13.16	3.14	100
	2000	75.25	6.80	13.87	4.09	100
Total OECD	*1981*	*66.12*	*15.18*	*16.08*	*2.63*	*100*
	2000	*69.46*	*10.38*	*17.21*	*2.96*	*100*

m: missing; x: included in another column
Source: OECD R&D database

What kind of influence will market forces have on post-secondary education in the future? Will they become prevalent, as it is now the case in many other activities formerly provided by the public sector? Will the university become more clearly demand-driven, leading to changes in its internal management and teaching practices as well as in a shift in its core missions? Increased use of cost-effective technologies and entry into foreign education markets are two responses of the university that are currently essayed in some countries: will these activities gain even more ground in the future?

Knowledge Economy

The expected decline in the university age population and the possible shift of public funding to other priorities might simply lead to a reduced size of higher education systems. This is an issue only because the production and transmission of knowledge has (and is expected to) become increasingly important for the growth of developed economies. Knowledge-intensive activities represent an increasing share of the GDP of developed economies. Similarly, the share of highly qualified jobs in total employment rose significantly over the last years, reflecting an increased demand for innovation activities (Foray, 2004). Although the high-level and often formal knowledge produced and taught in higher education institutions is just one kind of the required knowledge,[5] knowledge produced and transmitted in higher education is certainly a contribution to innovation, which is crucial in a knowledge economy. Building capacity for a knowledge economy thus requires a strong research sector and possibly the development of lifelong learning at an increased pace, within or outside the university. Again, those requirements could induce changes in the participation to and provision of higher education, including its research component.

The continuous development of Information and Communication Technologies (ICTs) is often considered as one of the drivers of the knowledge economy. While they have not triggered those developments, ICTs have probably contributed to their quick pace. The role of ICTs in higher education has already become important and could continue to grow in the coming decades. This is true for university teaching, research and management. Whether based on web, mobile or other technologies, e-learning is becoming a major kind of provision of distance learning and is increasingly complementing traditional face-to-face education. The University of Phoenix, the largest university in the United States in terms of enrolments, has for example 60,000 of its 140,000 students online. The Observatory on Borderless Higher Education undertook an international survey of online learning in Commonwealth Countries in 2002. Data from this survey reveal that of the 42 UK institutions that responded (out of a total population of 106), 62 per cent had developed an online learning strategy and most had done so since 2000. The main rationales for undertaking online learning were: enhancement of on-campus teaching and learning (first), improved flexibility of delivery for on-campus students (second), and keeping up with the competition (third). Enhancement of distance-learning was fifth (OBHE, 2002, p. 4).[6] ICTs are also playing an increasing role in research, allowing the quicker communication and collaboration of researchers across the world, but also increasingly the shared use of expensive resources, like computer power or simulation tools (NSF, 2003). While it is not clear whether these developments can lower the cost of teaching and research, they are opening new avenues for research and teaching.

Finally, the knowledge economy has led to more patent- and revenue-generating research and, in some cases, to more involvement of universities in local economies and industrial networks. For example, partnerships between industry and universities have increased significantly. As a producer of knowledge, the research carried out by university could get increased commercial value and capitalisation of knowledge could take increasing precedence over disinterestedness as a norm of science. This could lead intellectual property right issues to become more important in higher education, with for example the

---

5 A study showed that in the European Union, the most important level of education for regional growth was secondary education (and not tertiary education), probably because secondary education provides for the skills allowing for a strong organizational learning at the firm level (OECD, 2001).

6 A study on the uses, cost and benefits of e-learning as well as on the rationales and strategies for adopting it has been conducted by the OECD Centre for Educational Research and Innovation (0ECD, 2005).

patenting of knowledge becoming in conflict with its teaching. Intellectual property right issues between academics, universities, software designers, and others, may also arise as universities develop 'learning objects' that can be used in different contexts of e-learning and give this form of teaching an increased economic value.

These developments raise numerous questions for universities and higher education systems. Will the university carry out the main bulk of research or will revenue-generating research be mainly undertaken in R&D divisions of companies? Will the university provide skilled workers with the retraining that will become increasingly necessary in a knowledge economy? Will ICTs lead to a change in the provision of research and teaching? How will intellectual property rights regimes facilitate or hinder those developments? The increasing importance of knowledge for growth relies on transfer of knowledge as well as on innovation, giving partnerships between university and industry a big role for local economic development (Glasson, 2003). The knowledge economy thus gives universities more opportunities to play a predominant role in economic development.

New Actors in Higher Education

Some higher education institutions in several OECD countries have already started to anticipate and respond to these trends. New actors—corporate universities, consortia, virtual universities, and others—have entered higher education and have started to blur the usual borders between institutions: a phenomenon which is sometimes described as borderless higher education (Cunningham et al., 2000). Public universities increasingly operate for-profit; for-profit education providers increasingly offer degree-granting post-secondary education; commercial (and non-commercial) partnerships between universities, and between universities and industry, develop. While these developments are still marginal, they may point to possible directions for the future.

First, the borders between different types of educational institutions are no longer clear-cut. Higher education institutions no longer confine themselves to degree-granting education or private educational providers to professional training. Although they still represent a small share of the sector, for-profit or corporate providers are increasingly active and gaining ground in the degree-granting education sector. Traditionally, corporate training institutions provided vocational training for the staff of multinational companies of which they were often subsidiaries. The McDonald and Motorola universities are two examples. Now, one-quarter of these institutions are said to attract customers from outside the parent company (Meister, 1998; Cunningham et al., 2000). Across the world, Microsoft's 1700 Certified Technical Education Centres show how firms are moving into the vocational education market. These private centres operate as Microsoft franchises, with a training programme drawn up by Microsoft and taught by Microsoft-accredited staff. Internationally recognised by employers, these computer training programmes attract a large number of students (Adelmann, 2000). More and more of these corporate training institutions have appeared in recent years and have broadened their curriculum to attract a wider clientele with less company-specific courses. In the United States, their number quadrupled between 1988 and 1998, with 42 per cent offering courses that could have led to a diploma in accredited institutions (Densford, 1999). These institutions may thus increasingly become degree-granting institutions.

At the same time, expanding demand for professional and corporate training and language courses is addressed by an increasing variety of educational providers. In particular, traditional universities increasingly create private arms to address this demand, which has traditionally been considered as a private service in most (but not all) OECD countries. These trends are evident at the domestic level but also in cross-border education, where commercial activities have gained ground.

Second, partnerships of all kinds have boomed over the last decade in higher education. Partnerships between traditional universities or institutions have always existed for academic as well as for research purposes. Sometimes extended to management issues, they are more formalised today than in the past. Although academic partnerships have increased significantly in the last decades, they are not new. What is new is the inclusion of a greater number and variety of partners – educational institutions, providers and even corporations – and the shift to partnerships including commercial elements. These new cooperation arrangements may involve transfer of funds and have economic motives. A growing number of traditional universities have created consortia and partnerships to address new forms of demand (international demand, working adults, e-learning) through new modes of delivery. For example, Universitas 21 brings together 16 established universities from ten countries to pursue global initiatives that would be beyond their individual capabilities. In partnership with the company Thomson Learning, Universitas 21 has created Universitas 21 Global, an online (and television) provider, has started to deliver MBAs (Master of Business Administration programs) from 2003. Trium EMBA is another example of an international partnership, in this case involving US, UK and French universities delivering executive MBAs through a mix of face-to-face teaching and distance learning.

These partnerships respond to different challenges. First, e-learning requires a technological competence that is often better mastered by IT or media companies than by educational institutions. Second, partnerships can share the costs and risks of going cross-border, give more visibility to the institutions and take advantage of economies of scale. While they are still a small minority, an increasing number of hybrid organizations now operate in the global education sector. These new hybrid organizations have limited activity to date but may become more prominent.

SCENARIO BUILDING: GOALS AND METHODOLOGY

All these elements create an urgent need for the university to change, but leave scope for many possible directions. The different elements can affect the missions of universities in a variety of ways. In future, what could be the main missions of the university? What could be its sources of funding? How can the commitment of university to society in this rapidly evolving context be reinforced? How could these changes affect the learning/teaching process as well as the role and the position of the university in society? What would be the impacts for learners and academics, for the economy and the society? In another light, what kind of university will *society* need in twenty years? In times of rapid change, the possible future of the university needs careful scrutiny to create a common understanding of the challenges and opportunities faced by the university, and to help academia and education policy-makers to drive these changes in the most desirable direction in their specific context. This is where scenario building enters into play.

It is beyond the scope of this chapter to present the different goals pursued by and different kinds of methodologies used by foresight studies (Miller, 2003; Godet, 1991). This section will aim at clarifying some possible misunderstandings and highlight some limitations of the scenarios presented in the next section.

Although futures scenarios can try to forecast probable futures, for example by isolating different variables and weighting their impacts on key dimensions of higher education systems, this is not the aim of this scenario building exercise. Here, the scenarios are to a greater or lesser extent imaginative stories designed to facilitate forward-looking thinking, and ultimately strategic decision-making. While helping to cast light on the relationships between some parameters in higher education systems and engaging various stakeholders in an active discussion of the future of higher education, futures scenarios for universities help people to identify future desirable directions among a range of more or less probable possibilities. Foresight studies can thus have an immediate as well as longer-term impact on decision making. Knowing where one would like to go in the long run and how one's environment might evolve is indeed important even in short-term decision-making. Obviously, making wrong short term decision might hinder or impede the achievement of one's longer term goals. Moreover, it is important to look forward on the basis of a long term horizon because it leaves more choices for action. Waiting too long to deal with up-and-coming issues means the range of possible decisions often becomes much narrower than if action had been taken earlier on.

Scenario building involves several steps: identifying a system, its driving forces, its key dimensions, and then the selection of a limited number of contrasted and simple scenarios among numerous possibilities.

The System

In the present case, the system of study for the scenarios is the post-secondary education system, understood as a collection of specialised institutions and ways of carrying out several missions or functions for different groups of a population and for different kinds of knowledge: teaching, research, services to community, and credentialing (see below). Within this system, the focus will be on universities, understood as institutions carrying out the same functions and missions, but specialized in a specific kind of knowledge, general or vocational 'advanced' knowledge (that is, more advanced than at primary and secondary levels). Taking post-secondary education as the system to be explored allows to consider scenarios in which the specialization (or missions) of universities have changed — while continuing to assume that the missions and functions of post-secondary systems have to be performed, in one way or another. This allows for radical change in the organization of universities, like for example an extreme scenario in which universities have disappeared, leading to imagining of other ways the functions of universities could be performed. Post-secondary education is thus looked at from the perspective of universities; or, stating it otherwise, universities are seen as the entry point to post-secondary education systems.

The missions of universities are generally understood as teaching, research, and services to community. Teaching consists of transmitting a specific kind of knowledge – advanced knowledge building on basic and secondary education, rather than practical information or basic arithmetic – and training people (future researchers and academics) to produce or preserve (advanced) knowledge themselves, with a commitment to inquiry and truth. There is no need to define exactly what kind of knowledge this is: the point is that higher education does not deal with all kinds of knowledge. Research consists in producing new knowledge and in preserving or keeping alive (advanced) knowledge. Services to community refer to different possible activities that were usually not seen as the core activity of universities: lifelong learning (basically not different from teaching, but generally understood as teaching to an older audience), services to people (like medical care for university hospitals), social critique (a by-product of research), territorial development (a by-product of research), consulting (a by-product of research). Although the specificity of this 'services' mission is difficult to grasp, it highlights the requirements for research and teaching to be useful to the whole society, through tangible applications or contributions—and clearly indicates that universities have moved away from the 'ivory tower' model, or are asked to do so.

Less often acknowledged as such, the last function, if not mission, of universities lies in credentialing. The only remaining 'monopoly' of universities and, more generally formal higher education institutions, is the legitimate power to grant degrees. Therefore, universities play an important role in screening out abler persons for employers, education serving as a 'signal' of their potential human capital (Becker, 1964; Spence, 1973). Widely accepted as legitimate, these credentials are one of the foundations of social stratification in democratic societies—whatever the interpretation of the process (Foucault, 1975; Bourdieu, 1989). Hence, universities have four missions or functions: teaching, research, services and credentialing. The futures scenarios will endeavour to understand how these missions and functions can be performed in different contexts.

Key Dimensions and Selection of the Scenarios

The driving forces that may affect higher education in the future have already been highlighted in the first section. The key dimensions are: demographic changes; funding difficulties; internationalisation; technology; liberalization; new public governance; knowledge economy; increasing partnerships; diversity of providers; lifelong learning; intellectual property issues. A simple way to build scenarios out of all these variables is to select a few important dimensions, and explore the associated possibility space. Another way is to give a strong emphasis to one of these factors and shape scenarios by imagining its impact on the other variables.

Several sets of possibilities and methods have been explored. First, different combinations of the three main missions of universities (teaching, research, services) have been analyzed in different contexts (importance of lifelong learning, internationalisation, demographic trends) and, starting from the ending, scenarios which could lead to this outcome have been imagined. Second, scenarios have been generated starting from two key dimensions: the main funding mechanisms (market, quasi-market, state) and the main source of decision in (or control of) the definition of the higher education missions (students, business, state, academics). All possible combinations have been explored. Third, the dimensions have been crossed in another way to generate the scenarios: funding (state or market) and range of participation (19-25 years-old or lifelong learning). Fourth, scenarios have been generated by giving a strong emphasis to some dimensions (technology, market forces, demography, globalization). This has led us to imagine numerous stories, but the four sets of scenarios remain relatively close. Finally, the two following key dimensions have been selected: the range of recognized educational supply (degrees are controlled by a relatively small number of institutions or degrees are delivered by a large number of institutions covering a wide range of educational experiences) and the range of educational participation (mainly initial tertiary education or mainly lifelong learning). But the scenarios have kept some features that appeared more strongly during the exploratory phase.

Some scenarios are impossible because they include certain conflicting or inconsistent characteristics. While generating and exploring scenarios is useful to understand the dynamics and inter-relationships of the system (that is, what makes a scenario consistent), only consistent or possible scenarios are ultimately interesting for decision-making. However, one should not rule out scenarios on the mere ground that they are improbable. Among the possible scenarios, some can be more or less probable and/or desirable, and desirable scenarios may not be the most probable, and vice versa (Miller, 2003). Probable, improbable, desirable and undesirable scenarios can all be useful for forward-thinking. They can trigger forward-thinking and help determine what can actually be achieved. Actually, undesirable or improbable scenarios do generally trigger interesting thoughts on different aspects of higher education and help understand one's goals.

The proposed set of scenarios includes extrapolations of current trends as well as scenarios representing more radical changes compared to the current situation. More difficult than imagining scenarios for universities is the selection of interesting ones. There is indeed a host of possible scenarios for universities, among which one should select a small number of contrasting ones. All the difficulty is to find a good balance between stylized contrasted oversimplifications and a too large range of close-but-different scenarios. A second difficult aspect lies in the relevance of the scenarios in the current context. We can imagine that teaching will become useless, and replaced by brain implants; or that DNA will allow us to determine people's abilities and replace the credentialing activity of universities. This might provide good material for books or movies, but is not very useful for decision-makers interested in the future of higher education in 20-30 years. Indeed, their decisions could hardly influence such developments.

Finally, it is important to stress a limitation of this set of scenarios, namely its general level. The proposed scenarios are not specific to a national or regional context but remain at a common international (and thus general) level. Scenarios for a specific country (or region) would have led to different stories. In some countries, some scenarios may appear to be in the past while they may still be the future for others. However, history has shown that the imaginative recreation of the past could be a direction for the future. Moreover, the proposed scenarios are a general tool to reflect on different aspects of higher education and not a specific tool for understanding, say, the future of university financing, the future of academics, the future of university buildings, the future of university research, and so on. While the proposed scenarios integrate discussion on all these aspects of higher education, specific scenarios on these topics would probably be different. On the other hand, generality and international thinking has advantages: the scenarios draw on trends within the OECD area and are useful in understanding how similar goals can be achieved via different means (and which are sometimes already being experimented in a part of the world other than one's own).

SIX SCENARIOS FOR UNIVERSITIES

Today's stories about tomorrow inevitably face the fundamental constraints of language and uncertainty. The ideas and words that will be used in the future have not yet been invented or lived. Nor is it possible to know the 'facts' of a day that has not yet passed. As a result stories about the future are largely rooted in the present – the expectations, fears and hopes that form the path to the future.

In a similar fashion, the scenarios presented below are familiar, reflecting fairly closely the burning historical and institutional issues facing universities in OECD countries today. Figure 1 maps the six scenarios for universities. As mentioned above, the two key dimensions used to design and organize the scenarios are the range of recognized educational supply and the range of educational participation.

The range of educational participation is about the age composition of the student population. At one end of the axis, *initial education* refers to universities focusing on the initial education of young students coming directly from high-school. Like today, 80 per cent of university students would be under 23. At the other end, *lifelong learning* means adults attend university at different times of their life, possibly several times: initial tertiary education, re-training and leisure, and when they retire. Students are more diverse and, on average, older: possibly 50 per cent of them are aged over 40 years-old. The range of recognized educational supply is about the strength of the university monopoly over degrees. At one end, a restricted number of institutions grant degrees and have a say on the kind on knowledge that can lead to these degrees. At the other end, a variety of institutions grant degrees and a wide range of knowledge can thus be recognized by a degree.

Other variables have been emphasized to select six stories among the thousands in the possibility space. The selection and emphasis of the variables is designed to ensure internal consistency of the scenarios, but also to propose differentiated enough scenarios. The six variables selected for constructing the five scenarios are: 1) the type of population covered by tertiary education, as well as correlated variables; 2) funding (predominantly public, mixed, predominantly private); 3) the integration of missions offered; 4) the international dimension; 5) the homogeneity of status of faculty and institutions, and 6) the degree of take-up of technology (e-learning, e-research). Matrix 1 presents the six different stories in summary form.

Figure 1.1: Six possible scenarios for universities

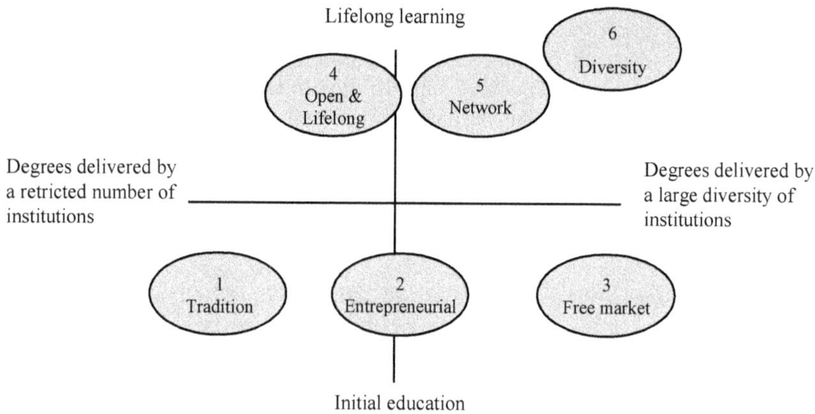

Scenario 1: Tradition. Universities are mostly like today, catering to a relatively small share of the youth population for the purposes of job selection credentials. Universities pursue both teaching and research, as now, without excessive dependence or involvement with the private sector. Governments continue, in most OECD countries, to play a prominent role in funding, regulating and managing universities. Within a public accountability and equity framework there is little scope for profit-generating initiatives. Lifelong and e-learning develop largely outside the university sphere.

Scenario 2: Entrepreneurial Universities. Selective institutions cater largely to young people in their initial preparation for life. The key difference with the previous scenario is that universities (public or private) can respond with greater autonomy to a variety of funding sources. There is a more mixed public-private funding model, with university resources coming from a wide variety of sources. Along with the returns to the intellectual property rights that it secures, research is seen as a very important and lucrative activity. However, universities take a market-oriented approach to operations without losing basic academic values.

Given the prestige and income accorded to research the teaching side remains quite elitist. As for lifelong learning, it occurs within a university setting but in teaching-only institutions with lower status. The three missions of the university – teaching, research and community service – are well balanced, though there is greater differentiation across institutions due to more autonomy and responsiveness. Commercial approaches to international markets and e-learning are important. University resources as well as wages and prestige of academic staff improve. Links to the local economy are strong.

Table 1.5. Matrix: Six scenarios for the future of universities

	Scenario:	*1*	*2*	*3*	*4*	*5*	*6*
1	a) Selective/Initial education/Mostly young students	X	X	X			
	b) Open/Lifelong learning/All ages				X	X	X
2	a) Mainly public funding	X					-
	b) Mixed funding		X		X		-
	c) Mainly private funding			X		X	-
3	a) Teaching & research ("+": with strong research)	X	X+				
	b) Mostly teaching				X	X	
	c) Specialisation by missions			X			X
4	a) Mostly national focus	X			X		
	b) Importance of international focus		X	X		X	X
5	a) Homogeneous status of staff and institutions	X			X		-
	b) Polarisation in status of staff and institutions		X	X		X	-
6	a) Low e-learning and role of ICTs	X					
	b) High e-learning and role of ICTs		X	X	X	X	X

- : undetermined

Scenario 3: Free Market. Market forces are the main drivers of this scenario. A private tertiary sector is regulated by private companies for quality assurance and accreditation and mostly funded through market mechanisms. Market forces give rise to institutions specialized by function (teaching, research), field (business, humanities), audience (young students, part-time students, distance education, adult education, lifelong learning) while business firms grant degrees to their employees for corporate training. Hierarchy between those very diverse institutions becomes very strong, with a global super-elite, and more polarization in the status of faculty. With the widening of student choice there is greater competition for students and tuition revenue comes to represent a more important share of overall income. Technology is widely used in teaching. The international dimension of the market is important. Since most students and their parents are not interested in research, refusing to bear the costs, research moves out to public research centres and corporate R&D. What research remains in universities becomes more elitist. Teaching to mass markets leads to greater standardization and the patenting of curricula and teaching methods. Research becomes more demand-driven, specialised and secures important returns through intellectual property.

Scenario 4: Lifelong Learning and Open Education. Universities are marked by universal access for all ages and much less research. The knowledge economy has flourished. Higher education provides for recurrent professional development financed by companies, individuals seeking recognized skill upgrading, and states. In an ageing society, more elderly people enrol for non-professional reasons. Universities become more learner- and demand-oriented, more teaching oriented, with short courses, more distance learning, and more e-learning.

Governments or independent accrediting bodies take charge of quality assurance and accreditation. Most research is done outside higher education with the best researchers moving to private companies, specialized institutes or the few remaining elite universities. Corporations and corporate universities have a large influence. Integration with the applied side of learning might go so far that all university education follows a professional school model. Responsiveness to market forces is high and there is much business oriented investment in e-learning.

Scenario 5: Global Network of Institutions. Post-secondary studies become demand- and mostly market-driven. The two main innovations are 1) that learners define their own course of study from across all available courses throughout the global post-secondary education network and design themselves their degrees; 2) that higher education institutions partner increasingly, including with industry. E-learning develops strongly in this scenario, as well as other means of education. The training content becomes more standardized and possibly embedded in technology and media (for example modular learning objects or edutainment through partnerships with game industry). The provision of and market for lifelong learning becomes very large, especially as education takes a multiplicity of new forms. Most research is carried out outside the higher education system, and faculty in mostly teaching institutions becomes less qualified than today but use more sophisticated teaching techniques. There is a strong polarization in the status of academic, with academic superstars and developers of 'learning tools' getting high status whereas the average teaching staff becomes less qualified and gets lower status. Programmes and courses matter more than institutions. Intellectual property rights for substance as well as teaching methods create high returns.

Scenario 6: Diversity of Recognized Learning – Disappearance of Universities. In this scenario, the formal tertiary education sector disappears. People learn throughout their life, at work, at home, for personal and professional motivations, more and more by themselves and by sharing their expertise with other people interested in the same field. Professional education requiring hands-on practice, like surgery, is transmitted within businesses through an apprenticeship system or thanks to new sophisticated electronic devices (for example online). Technology is an enabler for the diffusion of information and knowledge. People learn as much and possibly more than today but in a different way: learning takes the model of 'open course' education, mostly free and non commercial, involving a lot of partnerships between individuals and institutions of all sorts. Global networking is thus important and goes beyond institutions. Knowledge and experience acquired in all life situations are acknowledged through formal assessments of credentials carried out by specialized assessment bodies. But given its pervasiveness, knowledge is less of a determinant for a career or in the stratification of society. While research becomes less specialized in fields requiring little money, like humanities or mathematics, a large share of research requiring high investments takes place in public research centres and in corporate R&D divisions.

CONCLUDING REMARKS

It is important to look at all the scenarios with a fresh eye and to reflect on what their advantages and downsides may be. For example, scenario 1 may appear as the current situation in some OECD countries, for example in continental Europe, and also as what some developing countries are aiming to develop in the future. However, it is often considered that it has become obsolete with the process of massification. The future of the traditional university could thus be very gloomy (for academics at least): universities could continue to be under-funded and drift towards the secondary school model – with academics becoming more like school teachers. However, one may also argue that this traditional model is the future of the university, and will result from the development of different post-secondary institutions including institutions with smaller and elitist enrolments. The old ideal of the university to fuse teaching and research, by teaching to students that will themselves create new knowledge, could thus have a second life.

As a starting point, questions that may help in using the scenarios as a reflection or discussion tool: Where does one want to go? Which of these scenarios seems the most (and least) desirable? Why? Which seems the most (and least) likely? Why?

It is indeed important to understand the process that could lead from one point to the other. Stakeholders' strategies and interests must enter into play, to assess what is reasonably achievable or not. And the discussion should consider the impact of the different futures on core social values. For example, could the future university enhance equity, participation of under-represented groups and social cohesion?

Futures scenarios are not an end in themselves. They shall help to engage stakeholders in discussion and to tackle specific issues. While it would be tempting to comment on the scenarios and unveil their dynamics, drawing out additional details, discussing, criticizing and ultimately building one's own scenarios is the next step of the exercise – the final step being decision-making.

REFERENCES

Adelman, C. (2000). *A parallel postsecondary universe: the certification system in information technology*. Washington: Office of Educational Research and Improvement, U.S. Department of Education.

Altbach, P. (1998). Comparative perspectives on higher education for the twenty-first century, *Higher Education Policy*, 11, 347-356.

Becker, G. S. (1964). *Human capital*. Chicago: The University of Chicago Press.

Beerkens, H.J.J.G. (2004). Global opportunities and institutional embeddedness. Enschede: Centre for Higher Education Policy Studies, University of Twente.

Boer (De) H., J. Huisman, A. Klemperer, B. Van Der Meulen, G. Neave, H. Theisens & M. Van Der Wende (2002). *Academia in the 21st Century: An analysis of trends and perspectives in higher education and research*. The Hague: Adviesraad voor het Wetenschaps.

Bourdieu, P. (1989). *La Noblesse d'Etat. Grandes écoles et esprit de corps*. Paris: Editions de Minuit.

Braun, D. & Merrien, F.X. (1999). *Towards a new model of governance for universities? A comparative view*. London: Jessica Kingsley.

Cunningham, S., Ryan, Y., Stedman, L., Tapsall, S., Bagdon, S., Flew, T. & Coaldrake, P. (2000). *The business of borderless education*. Canberra: Australian Department of Education, Training and Youth Affairs.

Densford, L. (1999). Many CUs under development. *Corporate University Review*, 7(1).

DfES (Department for Education and Skills, United Kingdom) (2003). *The future of higher education*. Norwich:DfES.

Duderstadt, J. J. (2000). *A University for the 21st century*. Ann Arbor: University of Michigan Press.

Etzkowitz, H., Webster, A., Gebhardt, C. & Terra, B. R. C. (2000). The future of the university and the university of the future: evolution of ivory tower to entrepreneurial paradigm. *Research Policy*, 29, 313-330.

European Commission (2003). *The role of universities in the Europe of knowledge*. COM(2003) 58 final, Brussels: European Commission.

Foray, D. (2004). *The economics of knowledge*. Cambridge, MA.: The MIT Press.

Foucault, M. (1975). *Surveiller et punir*. Paris: Gallimard.

Glasson, J. (2003). The widening local and regional development. Impacts of the modern universities – a tale of two cities (and north-south perspectives). *Local Economy*, 18(1), 21-37.

Godet, M. (1991). *De l'anticipation à l'action*. Paris: Dunod.

Graham, G. (2002). *Universities. The recovery of an idea*. Thorverton: Imprint Academic.

Higher Education Authority / Sonas Innovation (2003). *The future of research and education networking in Ireland*. Dublin: Sonas Innovation.

Kerr, C. (1963). *The uses of the university*. Cambridge, MA: Harvard University Press.

Marginson, S. & Considine, M. (2000). *The enterprise university: Power, governance and reinvention in Australia*. Cambridge: Cambridge University Press.

Martin, B. R. (2002). The changing social contract for science and the evolution of the university. In A. Geuna, A.J. Salter & W. E. Steinmueller (Eds.), *Science and innovation: Rethinking the rationales for funding and governance*. Aldershot: Edward Elgar.

Maskell, D. & Robinson, I. (2001). The new idea of a university. Thorverton: Imprint Academic.

Meister, J. (1998). *Corporate universities: Lessons in building a world-class workforce*. New York: McGraw Hill.

Miller, R. (2004). University futures in a learning intensive society (mimeo).

National Science Foundation (USA) (2003). Revolutionizing science and engineering through cyberinfrastructure, D. Atkins (chair). Report available at: http://www.communitytechnology.org/nsf_ci_report/

Newman, F. (2000). The futures project: policy for higher education in a changing world. Available at: http://www.futuresproject.org/publlications/position_paper.pdf

Newman, J.H. (1854),.*The idea of a university*. Notre Dame: University of Notre Dame Press, 1982.

Observatory on Borderless Higher Education (2002). *Online learning in Commonwealth universities – Selected data from the 2002 Observatory survey*. Briefing Note. Available at: http://www.obhe.ac.uk.

Organization for Economic Cooperation and Development, OECD (2001). *Cities and regions in the new learning economy*. Paris: OECD.

OECD (2003). Changing patterns of governance in higher education. In OECD, *Education Policy Analysis*. Paris: OECD.

OECD (2004). *Internationalisation and trade: New challenges for higher education*. Paris: OECD.

OECD (2005). E-learning in post-secondary education. Paris: OECD.

OECD (2006). *Education at a glance*. Paris: OECD.

RENAUT, A. (2002). *Que faire des universités?* Paris: Bayard.

Scott, P. (2002). The future of general education in mass higher education systems. *Higher Education Policy*, 15, 61-75.

Spence, M. (1973). Job market signalling. *Quarterly Journal of Economics*, 87, 355-74.

World Bank (2002). *Constructing knowledge societies: New challenges for tertiary education*. Washington DC: World Bank.

Stéphan Vincent-Lancrin
Centre for Educational Research and Innovation,
Organization for Economic Cooperation and Development
Paris, France

SIMON MARGINSON

2. THE NEW HIGHER EDUCATION LANDSCAPE

Public and Private Goods, in Global/National/Local Settings

INTRODUCTION: WORLD-WIDE HIGHER EDUCATION

Higher education is situated in a transparent information environment in which words and images flow freely, borders are routinely crossed and the systems and institutions are continuously whirring. Arrangements and identities are conditioned by history and some appear more solid than others. But no project in higher education is its own final point of reference. The future has never been more open.

There has been a great change in the last two decades. It is about the global. We begin to imagine higher education as a world-wide convergence: not as a unitary 'global system' but a more complex combination of (1) global flows and networks of people and institutions, ideas, knowledges, technologies, finance; (2) national higher education systems shaped by tradition, law, policy and funding, and by equalities, competition and differentiation; and (3) single institutions operating at the same time in the local, national and global dimensions. Especially in research-intensive universities, the most globalized institutions, the ultimate horizon of perspective is the planet itself. Within that global horizon individual, institutional and systemic agency are continually being made and self-made, in encounters with diverse others and in reflexive negotiations with locality and with self.

If world-wide higher education is integrated it is an imperfect integration. There are ever-changing patterns of engagement; unevenness between sites in capacity and the use of common goods; zones of autonomy and more-or-less separation, and stable and unstable hierarchies. Relations are structured by both cooperation and competition. There are fecund mutual influences, doggedly persistent differences and often surprising similarities of approach across borders.

Bourdieu's theorisation of the 'field'. This bounded, fragmented, complex, contested, hierarchical, product-making, subject-forming, constantly transforming world-wide arrangement; with its malleable rules, discourses and exchanges; recalls Bourdieu's (1984; 1988; 1993; 1996) notion of a 'field of power'.

Bourdieu develops 'the notion of a field, understood as a space, that is, an ensemble of positions in a relationship of mutual exclusion' (1996, p. 232). A field of power is a social universe with its own laws of functioning. It enjoys a degree of autonomy, to the extent that it holds off external determinants and is driven by its own internal logic. Within the field agents compete for resources, status or other objects. There is inter-dependency between the prior positions of agents and the 'position-taking' strategies they select. 'Every position-taking is defined in relation to the space of possibles which is objectively realized as a problematic in the form

Simon Marginson (ed.), Prospects of Higher Education, 29–77.

of the actual or potential position-takings corresponding to the different positions' (Bourdieu, 1993, p. 30, his emphasis). Position-taking is the 'space of creative works' (Bourdieu, 1993, p. 39). This is not an open-ended free-wheeling creativity. Only some position-takings are possible, identified by agents as they respond to changes in the settings and the moves of others in the competition game. Agents have a number of possible 'trajectories', the succession of positions occupied by the same agent over time, and employ semi-instinctual 'strategies' to achieve them. Agents respond in terms of their 'habitus', their acquired mix of beliefs and capabilities, and in particular their 'disposition' which mediates the relationship between position and position-takings (pp. 61-73). Bourdieu asserts reciprocity between structure and agency. 'Although position helps to shape dispositions, the latter, in so far as they are the product of independent conditions, have an existence and an efficacy of their own and can help shape positions' (Bourdieu, 1993, p. 61). The scope for the self-determination of agents, 'the scope allowed for dispositions', is variable, being shaped by the autonomy of the field in relation to other fields, by the position of the agent in the field, and by the extent to which the position is a novel and emerging one, or path-dependency has been established (p. 72).

Bourdieu's theorisation of 'field' was not prepared for world-wide higher education and does not constitute a perfect analytical blueprint. Arguably there is insufficient space for agency and especially for conscious imagining and creativity in Bourdieu's conception. He sees freedom as mostly bound *a priori* by class power and resource distributions installed in the unconscious and expressed in instinctual responses in the competition game. This leads to an under-discussion of the potential for endogenous changes in the structure of positions in the field, which is mostly read as a reflection of economic and political forces outside the field. Bourdieu is too quick to universalise competitive behaviours and underplays cooperation, mutual assistance and the 'gift economy' (Kenway et al., 2006) which does not fit with his 'academic capital'. There is insufficient discussion about how knowledges are constituted (Naidoo, this volume) and function as tools of differentiation and power. Above all his field of power is nationally-bounded and misses the plasticity, mobility, volatility and disjunctiveness that characterize global relationships (Appadurai, 1996; Marginson, 2007b). Nevertheless, the notions of a bounded field of higher education with a relative autonomy in relation to external influences, and the interplay between position and position-taking, are illuminating and have helped to shape this chapter. Likewise Bourdieu's insights into the internal structure of the field of higher education, with its sub-fields of elite cultural production and mass commercial production; and his understanding of the credentialing function of education institutions and the scarce and differentiated cultural capital they create, have fed into the discussion of market competition.

This Approach of This Chapter

When surveying the complex terrain of world-wide higher education, any synthesis is partial. We choose a particular angle, or a selected and managed plurality of angles, from which to illuminate the whole. This chapter discusses higher education in terms of a new and emerging architecture of world-wide relations of power in the sector. It does so by drawing together two analytical frameworks, with origins in political philosophy and global sociology respectively. One framework is the distinction between public and private goods (Marginson, 2007a). The other is the distinction between global, national, and local dimensions of action (Marginson & Rhoades, 2002). Arguably, by synthesising these analytical elements we can more readily explore the potentials and limits of higher education in relation to both national systems and the global dimension at the same time, and grasp the transformative effects flowing between them. We can more clearly understand the setting in which each institution is placed, the changing scope for agency and self-determining action (the new field of freedom) in higher education, the new drivers and constraints, and the forms of differentiation and inequality that are produced. We can map relations of competition alongside relations of cooperation, finding that both are part of the mix: sometimes antagonistic to each other, sometimes not.

This last point provides an example of what the broader synthesis can do. In world-wide higher education we observe competition for status and power between research universities, competition for revenues between all institutions; and on the global plane, competition between institutions, and nations, for relative advantage and strategic position. In terms of cooperation, we observe freely exchanged knowledge flowing in the public domain, and networks of institutions, and research groups, and between individual collaborators. Competition and cooperation are different and opposing forms of human relation. Yet in practice the dividing lines between competition and cooperation, markets and non-market activity (and also economy and culture) are often crossed. Often we find that the opposing modes are both part of a package and their relationship is more symbiotic than antagonistic.

To anticipate the chapter's argument, one major part of the story is that despite the emphasis on commercial science in policy discourse, research mostly continues to be produced in the form of a free public good, *not* as a capitalist commodity in economic markets. In fact in this more global era the scope and influence of its production and dissemination as a public good have vastly expanded. Yet basic research is also a primary good in competition between elite universities. Research outputs are a key factor in defining the value of the private positional goods produced by institutions, particularly degree programs at the leading research intensive universities. In other words one of the fecund potentials of public good research is to provide a means of defining a hierarchy of institutions, and on the global plane, a hierarchy of countries, on the basis of research performance. In a world in which capitalism is the dominant economic system, and power relations are highly unequal within and between nations, the production of public knowledge goods has become annexed to the workings of competitive markets in higher education, and also the exercise of global hegemony by U.S. universities (and to a much lesser extent those of other English language nations) in higher education. In this manner one particular kind of public good research (Anglo-American led knowledge) is complicit in the crowding out of other kinds of public goods (for example in, say, Mexico, scholarship in the Spanish language); and in smoothing the way for the commodification of teaching in place of freely accessed degree programs (for example the growth of high fee private tuition in the private sector and the imposition of charges in place of free education in the public sector, all of which are supported by the emergence of a dominant U.S. model of the university).

So 'public good' is not as simple as it seems and economic commodities rest on non-economic conditions. This apparent category ambiguity is the consequence of using a more complex, multi-dimensional-sided analysis. In this process category distinctions such as public/private and global/national/local continue to be useful. Yet at the same time many practices in higher education refuse to stay confined to single categories for very long and some always had a foot in more than one camp. Category boundaries are continually being crossed and the practical terrain covered by each category is always shifting. Boundaries and coverage vary in place as well as time. The fact that such changes and changeability can be observed is a healthy sign. This means that the analytical framework is allowing a fuller reality to come into view. This is what happens when single category reductionism is left behind; for example those one dimensional narratives that pin all the causal blame on an abstract globalization, or pose the world of possibility in terms of either universal public goods or universal commodification. It does not mean that categories as such no longer work. It is more complexity than ambiguity, more the inability of binary categories to hold as mutually exclusive dualisms, than the explosion of category distinctions as such. After all, even if it is acknowledged that everything is historically relative, the fixed narrative certainties have gone and the future of 'The University' is wide open, there is still a need to reason and understand. And, fortunately, there are more theoretical and empirical tools (such as the data collected by international agencies) for doing this than once there were.

Nowhere is complexity and changeability more obvious than in the incessant activity on the borders between the global, national and local dimensions. Every practitioner in higher education is aware of these dimensions and knows they have different mediums of operation and points of reference. Everyone makes category separations between the three dimensions on a daily basis. And everyone finds that each of the dimensions is continually being pushed against and into the others.

Sequence of argument. The chapter begins by setting out the public/private and the global/national/local distinctions and briefly explores globalization in the sector. It then explores the workings of higher education markets, especially global markets, whose role is advancing rapidly on the world scale, along the way exploring also public goods in research and knowledge. (This chapter outlines the role of research-based differentiation within the larger map of power; chapter 3 explores global research and university rankings in more detail). The chapter reflects also on the wash-back effects in national systems of global competition and public goods.

Inevitably the ambitious scope of the chapter has collided with the constraints of space and it is best understood as an invitation to reflection rather than a definitive analysis. Some elements, such as the new potentials of individual and institutional agency in higher education, the cross-border practices of institutions, the character of global flows, the evolution of national and regional roles in the global setting, including Europeanisation, and inequalities between and within national systems, are merely sketched or are undiscussed here. Some are developed elsewhere (for example Marginson 2007b; forthcoming; Marginson & Ordorika, 2007; Marginson & Sawir, 2005; 2006; Marginson & van der Wende, 2007; work in progress).

In this chapter higher education is discussed mostly in relation to the teaching, research, servicing and administrative activities of research universities as well as those research institutes that specialise in parallel forms of basic research.

PUBLIC AND PRIVATE GOODS IN HIGHER EDUCATION

Higher education produces multiple and heterogeneous public and private goods. These goods are accessible to empirical observation and expert judgement, and in some respects to numerical measurement. Public and private goods will be defined and briefly discussed in turn, with some remarks about the relations between them.

Definition of Public and Private Goods

There are two conflicting approaches to the distinction between public and private goods. The first notion is the juridical definition of public/private. The boundary is determined by legal ownership. 'Public' goods are produced in the government or state sector. The notion is long-standing but is not used here, for several reasons. A definition based on juridical ownership tells us nothing about the character of the goods: for example accessibility, reach and social effects, whether they constitute individual or joint benefits and how they are financed. It is ambiguous as to the role of government, which often is expressed through the regulation and simulation of market and private sector activities; and ambiguous as to the definition of 'private' which can be variously understood as the commercial sector, the domain of family, or the imagined space that constitutes personal identity. Most crucially this definition negates the possibility of global public goods. Where 'public' is defined to mean state or government sector, 'in the international sphere, where there is no government, how are public goods produced?' (Kaul et al., 1999, 12). So how then can we discuss common international benefits and cross-border effects between one national system and another? It becomes impossible to understand global effects as anything other than 'private' and by inference, 'market', effects; and only the nation-state is capable of producing 'public' goods.

The economic definition of public/private offers a more promising approach. Following Samuelson (1954), public/private is determined by the character of the goods, in particular by whether they can be sustained by market production. Goods capable of full market production are defined as 'private'. All other goods, which will tend to be under-produced in unregulated markets, require state financing and/or provision and are at least in some respect 'public' in character. Samuelson defines public goods (or services)[7] as goods that are non-rivalrous and non-

[7] In this chapter the term 'goods' is used in a generic sense to refer to all forms of production including those industries conventionally characterised as 'services' such as education. 'Goods' refers to benefits obtained, which includes benefits that are intangible/ non-corporeal, as well as those manifest in corporeal commodities.

excludable. Goods are non-rivalrous when they can be consumed by any number of people without being depleted, for example knowledge of a mathematical theorem. Goods are non-excludable when the benefits cannot be confined to individual buyers, such as social tolerance, or law and order. Few goods are both fully non-rivalrous and fully non-excludable but many have one or the other quality in part. Goods with neither quality are classified as fully private goods. Some public goods take the form of 'externalities' or 'spill-overs', whereby an individualised good received by one person creates benefits for others that they did not purchase. For example the training of a technician in the workplace can enhance the productivity and wages of other workers; the training is partly non-excludable. Other public goods include collective benefits, for example the benefits created by enhanced communication or knowledge systems, where the benefits are non-rivalrous.

Samuelson's definition is rarely used precisely in the economics of education. Public goods are underplayed, in part because they are not always accessible to quantity measures (Marginson, 1997). Nevertheless this definition is a useful starting point for analysing goods produced in capitalist economies. In summary:

> Public goods in higher education as goods that (1) have a significant element of non-rivalry and/or non-excludability, and (2) are goods that are made broadly available across populations; and are inter-generational in that they meet needs in the present generation without jeopardising future generations. Goods without attributes (1) and (2) are private goods.

However, when utilising the public/private distinction as a tool for the *political* analysis of higher education and other complex services, the conception developed in economic theory can only be applied with two caveats. These caveats shift the public/private notion some distance from the foundations of economic theory.

The first caveat goes to the social construction of higher education. Mainstream economic theory is ahistorical. It seeks to fix its descriptors in a timeless nature. As Samuelson sees it, the categories 'public' and 'private' are intrinsic to the character of the good. He considers most goods to be 'naturally' private and accessible to full market production. But there is nothing intrinsic about human needs for complex cultural and economic goods. Higher education, like, say, communications and health, can be organized either predominantly as public goods or as private goods. Whether universities are public, in the sense of producing non-rivalrous or non-excludable goods under-produced in markets, is determined not by nature but by public policy and social practices. Universities can be free, open to all, with little or no status hierarchy between institutions and focused on research designed to solve problems such as ecological instability. Or universities can be costly, closed, stratified and focused on privately valuable degrees and technologies sold to the highest bidder. A third possibility is a mixture of public and private goods. The public/private character of higher education is open. Higher education is policy determined,[8] in that governments can enhance its 'privateness' or 'publicness', it is

[8] Often economists attempt to develop economic and policy analyses of education on the basis that it is intrinsically public or private or a fixed kind of intermediate case. For example some economists

nested in culture (Calhoun, 1998), and varies by time and place. There are markedly different configurations of higher education around the world. In short, the nature of the goods does not determine the character of production. The character of production determines the nature of the goods.

The second caveat is related and goes to the idea of public/private as a dualistic and unitary set as required for orthodox economic calculation (Dow, 1990). As the above suggests, complex 'products' like higher education are meta-outputs. They have multiple character and may carry many elements of both public and private goods. The same 'output' (a degree, a research inquiry) might have elements of both public and private character as Samuelson describes them. In other words, in contradiction to mainstream econometrics, relations between public and private goods in higher education do not have to be zero-sum. These qualities are sometimes but not always mutually exclusive. Much depends on how higher education is organized, financed and distributed; in other words, on the social construction of the goods in question. Under certain circumstances one kind of good grows at the expense of the other. For example, when open admission to an unlimited number of free university places is replaced by quota limits and high tuition, new private goods produced by higher education are created while the public goods diminish. Under other circumstances public and private benefits increase together. For example, growth in the number of places in higher education offering private benefits to individuals may expand spill-overs to other individuals and become associated with the collective benefits flowing from a general rise in the threshold of community and workforce literacy. In this instance private and public goods are positive sum. In fact, public and private goods are often inter-dependent, in that the production of one kind of good provides conditions that enhance the potential for the other. For example, an expanded basic research system provides positive conditions for the growth of commercial research.

It is possible for state-owned institutions to produce private goods, and privately owned and for-profit market institutions to produce public goods; though for the latter public goods are incidental to their main purpose. Unlike non-profit institutions, for-profit institutions are unlikely to focus on the production of public goods as an end in itself. Generally, where higher education is reorganized into a competitive economic market with high tuition, so that the role of private goods is enhanced while public goods such as social equity diminish, public/private relations become more zero-sum. Whether and to the extent that public and private goods are inter-dependent and feed into each other, or are mutually exclusive, is - like the public/private boundary itself - sensitive to policy and material limits.

---

argue that education is a 'club good', meaning that is non-rivalrous in consumption but is excludable, like a film screening (Kaul et al., 1999, p. 509). However the concept of education as a club good does not do justice to the historically variable character and also the multiple character of higher education. For example basic research is not excludable, or at least not for very long. Education is potentially rivalrous or non-rivalrous, excludable or non-excludable.

Public and Private Goods in Higher Education

Public goods produced in higher education include on one hand the 'spill-overs' derived from the education of each students; on the other hand collective goods such as its contributions to knowledge via the production, codification and research and scholarship; the constitution of common literacy and culture; and the formation of attributes of sociability such as cosmopolitan tolerance. Stiglitz (1999, p. 308) notes that knowledge is about as close as possible to a 'natural' public good. The mathematical theorem retains its value no matter how many times it is used or how many people use it. Nor can its benefits long be confined to particular individuals. Knowledge become a temporary private good via intellectual property regimes, but does not stay so confined, especially in a networked environment. Non-rivalrous and only temporarily excludable, it is always under-produced in markets. Here cultural attributes differ from knowledge. They can take the form either of shared public goods or individualised private goods as in Bourdieu's cultural capital.

Although the social opportunities allocated in higher education often take the form of private goods distributed under conditions of competition, the function of social allocation (whatever the outcome) is itself a collective public good, and it can be tweaked so as to expand other public goods, such as the more democratic allocation of opportunity. Equal educational opportunity without regard to private finances is a good always underprovided in markets. This is a principal driver of state regulation, financing and provision of higher education around the world.[9]

The most potent private goods produced in higher education are individualised status benefits or 'positional goods' (Hirsch, 1976; Marginson 1997). Institutions allocate scarce places that provide students selected into those places, and/or able to purchase those places, with immediate prestige, and with the opportunity post-graduation to secure superior incomes, social status and the 'social capital' derived from student networking. The positional good is eventually enfolded in the degree certificate, a portable currency of the claim to social position.[10] The value of the

---

[9] Nevertheless, unless the total number of places is increased, programs that create more places for some students also subtract places from other students. Affirmative action programs have both a common public good aspect (they contribute to fairness) and a private good aspect subject to rivalry and excludability (access to scarce university places). In the USA there have been intense debates around these issues, for example in relation to the University of California system (Pusser, 2003). In themselves conceptions of public/private goods cannot solve distributional policy issues. However, they can contribute to policy frameworks in which the issues are identified, negotiated and resolved.

[10] Arguably, both Bourdieu's 'cultural capital' and the economists' 'human capital' are less explanatory of education markets than is the notion of positional goods. Bourdieu emphasises the qualities carried by individuals, with less regard to the social demand for those qualities. Although he is interested in problems of changing social valuation, arguably the concept of 'cultural capital' contains an implicit bias to the aristocratic. The economist makes the opposite error, calculating the value of 'human capital' in terms of its take-up in the occupational marketplace. This interprets social demand and the utility of credentials too narrowly, confining this to the monetary earnings benefits without regard for status or psychic rewards (for more discussion see Marginson, 1993; 1997). In other words each theorization has a one-sided perspective on the social allocation function. Grasping social allocation from both sides is the key to the unity of production, consumption and distribution.

status benefit depends on the prestige of the institution. These positional goods are arranged in a hierarchy of value, so that high prestige universities allocate the highest value status goods. Such goods are often but not always produced and distributed in a competitive market of institutions (Frank & Cook, 1995; Winston, 2003; Geiger, 2004), depending on the national arrangements. When access is mediated by both scarcity and tuition prices 'status within groups emerges as a good like any other that is traded in the market place' (Frank, 1985, p. 10).

Higher education confers other private benefits, such as those related to personal formation and cultural acquisition as ends in themselves. But arguably, only positional goods can constitute the basis of competitive markets, in relation to the teaching and learning function. Positional goods are readily arranged in a hierarchy of value of consumption. In turn the hierarchy of consumption values is grounded in and reproductive of the hierarchy of producer institutions. The constitution of the status of positional goods in higher education, and the constitution of the status of universities as institutions, are discussed further below.

Global public and private goods. The global dimension of higher education is also discussed further below. For the moment it will be noted that global public goods are defined as follows:

> Global public goods are goods that have a significant element of non-rivalry and/or non-excludability and made broadly available across populations on a global scale. They affect more than one group of countries, are broadly available within countries, and are inter-generational; that is, they meet needs in the present generation without jeopardising future generations (Kaul et al., 1999, pp. 2-3).

Global public goods include collective global goods, and positive or negative global externalities. Collective global goods are obtained by nations and/or institutions from cross-border systems common to the world or a meta-national region, for example via regulation, systems and protocols; such as the Washington Accords in Engineering, and the Bologna Declaration of a common European higher education space. Global externalities arise when higher education in one nation affects significant numbers of institutions in other nations; either for better, for example the transfer of research knowledge; or worse, for example the 'brain drain' of talented people. Negative externalities are known as public 'bads'. Both markets and government can constitute global public 'bads'. Global public goods, like all public goods, tend to be under-provided in markets.

Once the global dimension is introduced, the picture becomes more complex than in the nation alone. Potential goods occur in a multitude of sites. Public goods can be common across different nations or specific to one. New cross-border and global agents come into the picture. International agencies such as the World Bank, and multilateral forums, can directly create global public goods/'bads', particularly collective goods. Another complication is that something constituting a global public good in one nation, may be a global public 'bad' in another nation. Global public goods, and their ambiguities, are further discussed at the end of the chapter.

Global private goods are neither non-rivalrous nor non-excludable, are subject to the transfer of benefits across national borders, and have value in more than one nation. In higher education one such set of global private goods is generated in commercial research and intellectual property. However the principal global private goods are degrees obtained either by students crossing national borders into the exporting nation, or programs or providers crossing national borders into the nation importing education. Foreign degrees are global goods in two senses. They are goods that are obtained in border-crossing, and they can be utilised in more than one nation. Global private goods are also explored further below.

GLOBAL, NATIONAL, LOCAL

A 'Glonacal' Framework

In the world-wide setting, higher education is manifest simultaneously in three dimensions: the global, the national and the local.

Here the term '*global* dimension' refers to the world level, a planetary spatiality. 'Global' refers to spaces, systems, relations, elements, agents and identities constituting and constituted by the world as a whole, or by large parts in pan-national regions such as Europe. It includes 'global flows' (Appadurai, 1996) in communications, transport and financial systems, and elements such as language and research exchange that integrate nations and agents across borders. 'Global' as used here rests on a particular configuration of general/particular. The 'global' dimension does *not* mean total or universal. It does necessarily include every national and local element, only elements that are part of the constitution of the world as an integrated world.[11] Thus 'globalization' in higher education and research refers to the making or the enhancement of these global spheres of action including global spaces, systems, elements, agents, identities and practices.

This includes perspectives or ways of seeing in higher education, which are more often global than they were. National policy and institutional performance are increasingly likely to be referenced to the requirements and measures of global examples and standards. This global referencing has been encouraged by the partial liberalisation of international trade in education, global university rankings, and world-wide science, research and academic publication; as well as global parallels and convergences in policy and approaches to management. Amid the first open information environment in history, national policy and culture can only partly insulate national systems from the relentless pressures of comparison and imitation in the performance-driven higher education sector. Nonetheless the global dimension is not all-consuming, nor are global effects uniform everywhere. Many institutions have no obvious global connections and agendas and are benchmarked locally. In any case global transformations are subject to complex variations, and articulated through the differing national and local zones. Relations between the

[11] The term 'world-wide' can be used for the totalising inclusive concept that takes in every global, national and local element.

national, local and global dimensions are historically relative, place relative and a case by case matter (Deem 2001; Marginson & Sawir, 2005). The quality of global engagement varies from nation to nation. The global flows between different nations, and between institutions across borders, can be two-way or uni-directional. These flows may augment the position of one party, or they may augment both.

In the global setting individual institutions are in continuous contact with each other, a growing number source part of their revenues from across national borders, and many have been granted greater autonomy in relation to national governments, in part so as to secure global effectiveness. Even as the global dimension is developing enhanced potentials, so institution-specific identity and *local* practices are developing enhanced potentials. At the same time *national* policy and resources continue to be important. With single exception of some virtual 'e-universities', and notwithstanding a partial tendency to 'disembedding' of research-intensive from national policy (Beerkens, 2004; Marginson & van der Wende, 2007) institutions of higher education and research remain at least partly grounded in the national setting and affected by policy, funding and regulation. The difference is that the nation is no longer the horizon of vision or possibility in higher education. The national project remains salient in the sector, and is the source of much of the dynamism of newly emerging science powers such as China and Singapore, but is now a globally referenced project. For governments higher education and research are primary instruments of national competitiveness in the global setting. It is the desire for global effectiveness that drives the emerging systems and universities.

These mutually constituting global, national and local elements interact continually, without any one dimension determining the others on a permanent basis. Marginson & Rhoades (2002) propose a 'glonacal' analytical heuristic[12] for understanding the global, national and local elements shaping higher education:

glonacal = *glo*bal + *na*tional + lo*cal*

Global Fusion of Private and Public Goods

Globalization consists of engagement, integration and convergence on the world scale, the 'transformation in the organization of human affairs by linking together and expanding human activity across regions and continents' (Held & McGrew, 2000, p. 54). Held and colleagues define globalization in short as 'the widening, deepening and speeding up of world wide interconnectedness' (Held et al., 1999, p. 2). They develop a more detailed definition specifying global transformations in space/time (see also Harvey 1990; 2006). As globalization advances cross-border interactions become more extensive, intensified, regularized and faster. Local and global dimensions are more intermeshed: local events can be transmitted everywhere and distant events have a magnified impact. 'Globalization' is:

---

[12] For one application of this glonacal analytical heuristic see the comparison of the global strategies of national universities in Indonesia and Australia by Marginson & Sawir (2006).

> A process (or set of processes) which embodies a transformation in the spatial organization of social relations and transactions – assessed in terms of their extensity, intensity, velocity and impact – generating transcontinental or regional flows and networks of activity, interaction, and the exercise of power (Held & McGrew, 2000, p. 55).

While international relations across borders might involve just two nations (inter-national) globalization takes in many nations. It is a dynamic process that draws the local, national and global dimensions more closely together.

Though contemporary globalization is often explained in terms of single causes (market forces, neo-liberal government, American imperialism, communications technologies, knowledge-intensive production, etc.), it has more than one root; and as its variation by time and place suggests, it is also articulated by context. In many countries the context of the accelerated globalization of the last 15-20 years has been dominated by ascendant neo-liberalism in government and organizational culture (Marginson, 1997). This originated in the early 1980s, predating the 1990s Internet and the turn to the global, and has coloured the contents of globally transmitted messages and policies. The policy momentum for market liberalization in trade, signified by the neo-liberal rhetoric of World Trade Organization, has led many to conclude that globalization is simply a neo-liberal project. The fact that global people flows in higher education are partly manifest within a buyer-seller market encourages the 'common sense' perception that globalization in higher education is simply policy-driven capitalism writ large. But arguably, without neo-liberal policy, the Internet and the accelerated people mobility would still occur, though perhaps with different forms and contents (Rhoads & Torres, 2006).

At its roots globalization is a symbiosis of economic changes and cultural changes. It is also a symbiosis of public and private goods. On one hand it rests on the formation of world-wide markets creating private goods, operating in real time via automated processes and underpinned by the first world-wide system of financial exchange; and growth rates of foreign direct investment that far exceed capitalist growth as a whole. With instantaneous transmission of financial information the turnover time of economic capital tends towards zero (Mandel 1975); the world economy moves faster and global operations become more transformative of the localized parts. On the other hand, globalization rests on new worldwide systems of communications, information, culture and knowledge. These cultural systems, which are partly subsidized by governments as public goods (for example universities, especially in basic research) have been mobilized by nation states and global agencies so as to support the extension of and operations of the global economic markets producing private goods. In turn global economic forces extend and intensify cultural integration and the world tends towards a single cultural community as McLuhan (1964) predicted.

Instantaneous communications and meetings, along with complex data transfer, are implicated in most other kinds of global transformation. The Internet facilitates the evolution of unitary sets of knowledge. Information and communications technologies (ICTs) bring world-wide voices, data and knowledges to each locality with incalculable effects on the imaginings, options and identities of local agents.

Information and communications are sustaining new forms of continuously engaged inter-subjective human communities, of unprecedented scale, variety and flexibility. These technologies are at the core of global economic and global cultural change, and the junction between them. It is this element above all which constitutes what is new about contemporary globalization.

Globalization and higher education. Higher education and research are among the most globalized of sectors and enmeshed in all aspects of globalization. They are foundational to the sophisticated use of technologies and to culturally complex communities. Like ICTs higher education is formative of the emerging global environment. Research universities are intensively linked within and between the major global cities that constitute the nodes of a globally networked world. Castells (2001, p. 225) remarks that the Internet 'allows metropolitan concentration and global networking to proceed simultaneously. The networked economy, tooled by the Internet, is an economy made up of very large, interconnected metropolitan regions'. Those regions typically exhibit a high density of higher education. Correspondingly, nations and regions relatively decoupled from the globally networked economy typically have a low density of higher education.

Higher education and research are primary elements in the capacity to be pro-active across national borders. Being implicated in global transformations, higher education is itself transformed on both sides of the economy/culture symbiosis. It is swept up in global marketisation: it trains and resources the executives and technicians of global businesses; the main areas of student growth are in business, computing and English language, the most globally mobile qualifications to have; and in higher education itself the first global markets have emerged, as will be discussed. Arguably, larger changes are happening on the cultural side, including communications, research and knowledge. Teichler (2004) remarks that 'it is surprising to note how much the debate on global phenomena in higher education suddenly focuses on marketisation, competition and management in higher education. Other terms, such as knowledge society, global village, global understanding or global learning, are hardly taken into consideration' (Teichler 2004, p. 23). This is ironic, given that while higher education is a second level player in the circuits of capital and the direct creation of economic wealth, it is pivotal to research and knowledge systems; in linguistic and information systems; and has thickening connections with media and communications. In a networked higher education world the potential intellectual communities multiply continually.

However, it is important to remember that national systems and institutions do not participate in world-wide higher education on the basis of equality. The length of time that more than 90 per cent of the population is enrolled in education varies from 15 years in Belgium to one year in India. The USA spends about $360 billion per year on tertiary education (OECD, 2006, Table 2.6 below) while some nations spend less than $10 million. The distribution of the competences needed to enable institutions and national systems to operate as self-determining agents in the national and global dimensions – journal access, scientific equipment, trained people, English language, stable institutions, communications infrastructure, modernized administration, executive steering, competitive faculty salaries, the financing of student support, and so on - is highly uneven. Globalization draws nations into common systems but excludes most of them from global power.

GLOBAL MARKETS IN DEGREES AND KNOWLEDGES

Roots of Markets in Higher Education

The potential for economic markets in higher education rests on the historical and political conditions. There is nothing inevitable about these markets. For the most part education markets in national systems are constructed and managed by governments, which stratify institutions and install relations of competition, prices and economic incentives (Marginson, 1997). In this, the neo-liberal era in policy and government, the development of the market form, including fully commercial higher education (Breneman et al., 2007), has been much advanced. Partial marketisation is a feature of many if not most national systems (Teixeira et al., 2004). In the global dimension, where there is no state, the potential for economic markets derives on one hand from national-level changes to facilitate cross-border trade, on the other hand from the enhanced global flows of people, knowledge, ideas and educational capital. But economic markets and commodities in higher education have not been entirely imposed from outside, whether by neo-liberal policy or by global convergence. They are also grounded in higher education itself.

Markets in higher education, and the stratification of the value of products and producer institutions, rest on two core functions of research intensive universities unique to those institutions. Interestingly, these unique functions do not include teaching, which is carried out also in schools, vocational training and almost every other kind of educational institution; nor occupational preparation which is shared with non higher education institutions, corporations and the workplace itself. The two unique functions are (1) advanced educational credentialing that mediates upper level careers, whether occupationally specific or not, that is, the formation of positional goods; and (2) the formation of knowledge via basic research and advanced scholarship. Knowledge production is not confined to higher education institutions; it is shared with a growing number of institutes, companies, R&D start-ups and self-employed experts. However, basic research and research training are largely controlled by the research universities and basic research institutes.

Markets in higher education are markets in positional goods. Highly sought-after universities have a scarcity of places. Valid applications exceed student places available, signifying their positional value. The *US News and World Report* (2007), which frames the competitive market in American higher education, uses student selectivity as one component of its index. Individual consumers express their choices in positional terms. But something else underlies those choices. In fact different levels of student demand and scarcity of places alone *cannot* define the hierarchy of value, or the market would be premised on a tautology: high positional value generates high positional value. There has to be a 'gold 'standard', external to the positional goods themselves, that defines value in the positional market.

Inherited status distinctions play a role, but are not decisive: they must be sustained by contemporary drivers of reputation. The 'gold standard', at once both traditional and modern, is comparative performance in research and scholarship. Research and scholarship are largely non market goods but accessible to various output measures and/or informed judgements. We know that the world's elite universities are elite not because of the earning power of their degrees (which is rarely measured) or the networking benefits of attendance (real, but subterranean to the public gaze) but because they produce most of the world's basic research and award the most prestigious doctoral degrees. Research output, and the reputation it generates, function de facto as the generic standard of 'quality' in higher education.

High quality research universities produce what are seen as high quality degrees. This, not the episodic role of research in pedagogies, is the teaching/research nexus foundational to higher education.[13] Markets in higher education are a symbiosis of economic elements conditioned by cultural elements, similar to globalization as described above. The production of private positional goods is underpinned by the production of public knowledge goods. Basic research with its open-ended social potentials is harnessed to the noble missions of status accumulation, product differentiation and competition gaming. Much of the creative potential of research and scholarship remains but it is shaped and limited in material terms by the role of knowledge as measure of status. Scarcity of status is mirrored by the concentration of research. Even at the other end of the market, where institutions rely on teaching and commercial service, focus on student numbers and revenues and ignore knowledge formation, their lesser status is defined by the absence of research. With the exception of selective islands of occupational training in certain countries, higher education without the kudos of research tends to have low positional value.

Positional Competition in Higher Education

Positional competition in higher education is more widespread than the presence of economic mechanisms such as prices. It turns on the comparative reputation of institutions. Whether the national culture emphasises capitalism or social egalitarianism, as long as there is an inequality between institutions in resources, functions and social regard, so that it is possible to secure a private advantage by attending one institution rather than another (that is, as long as there is a hierarchy of value in private goods), positional competition will be found. The positional competition is more or less intense depending on the steepness of the hierarchy between institutions and the scope for choice-making.

Prospective students do not always have a choice of institution. But they do choose in most nations, and where they do the positional aspect - though it is not the only consideration - is normally important. It is usually more important than teaching quality. Institutional reputation is known, teaching quality mostly is not. The acid test is that when faced by choice between a prestigious university with

[13] There is little clear evidence that there are consistent positive links between research work and the quality of undergraduate teaching, and vice versa.

known indifference to undergraduate teaching, and a lesser institution offering better classroom support, almost all students opt for prestige. This does not sit well with the 'student-centred' pieties of quality assurance but it is confirmed by research. Moogan et al. (1999) find UK students are more influenced by university prestige than measures of program quality. In Australia, when discussing factors influencing the choices of prospective undergraduates, James et al. (1999, p. xvi) note that 'applicants focus on broadly conceived course and institutional reputations when making their selections'; and that 'course entry scores, and by implication "university scores", serve as a proxy for quality in prospective students' eyes'. Applicants were making individualised choices, and saw higher education as a competitive market. But they did not focus much on teaching, or the lifetime rates of return imagined by human capital theory. Something older was at work, based on instinct not calculation: relative advantage. Relative advantage was grounded in comparative reputation, with student selectivity as its proxy.

In *Social limits to growth* (1976) Hirsch analyses the dynamics of positional markets. The related argument by Frank (1985; 2001) on 'winner-take-all' markets discusses status competition in American higher education, as does Geiger (2004). Whether or not full price tuition is charged, positional competition does not function in the manner of a conventional economic market. Hirsch emphasises the zero-sum character of positional competition. Elite degrees and other positional goods confer advantages on some by denying them to others. 'What winners win, losers lose' (Hirsch 1976, p. 52). Within a single higher education system there is an absolute limit on the number of positional goods at a given level of value. The number of such goods cannot expand without reducing unit value. For example if everyone can enrol in Medicine and become a doctor, Medicine ceases to be a high income high status profession. Given the absolute limit on the number of high value positional goods, there is an absolute limit on the number of high value institutions, and on the size of individual institutions within the prestige group. Elite doctoral universities cannot expand their production to meet full potential demand, becoming ever larger, like a Sony or a Starbucks, without cruelling their raison d'etre. They need revenues, and arrange their tuition regimes in profitable configurations, maximise philanthropic and research funding, and sell non-core services. But revenues are a means to the real end, which is academic and social status, signified by consumer preferment and underpinned by research reputation. In tuition-based systems such as United States' higher education, reputation is confirmed by a 'sticker price' that capitalises status as well as value:

> Prestige ought to reflect quality, but far more is involved. As a function of consumer awareness, prestige is affected by the entire manner in which selective institutions market themselves and how they are treated in the media. Specifically, rankings advance their own definition of prestige, creating a 'positional market'... The positional markers in this competition ... are measures of selectivity, costs, or rank (Geiger, 2004).

In a positional market there is both competition between producers and competition between consumers. Producer universities compete for the custom of preferred 'customers', students with the highest entry scores. Student 'customers' compete for entry to preferred institutions. Prestige sustains high student scores, competition drives them higher, and their very scarcity reproduces the prestige of the elite universities, in the kind of circular effect that always drives the reproduction of hierarchy. Positional markets in higher education are a matching game in which the hierarchy of students/ families is synchronised with the hierarchy of institutions. The peak group in each hierarchy is steeped in the habit of sustaining the other. Wealth follows prestige. Bright students are mostly also students from affluent and powerful families. Wealthy families invest in high value education to maintain their own social and professional leadership; and in so doing they sustain the universities that educate them as engines of status for both family and institution.

At the peak of the positional hierarchy research performance and postional advantage feed each other. Again, under the sign of status we find the characteristic inter-dependence between status generation and revenue generation. High research performing universities with leading academic staff attract bright students and their (mostly affluent) families. These institutions accumulate prestige which is cashed out as tuition revenues and further leveraged to raise public and private monies that buy high-cost faculty and sustain research programs. Measurable research performance (publications, citations, grants, star professors) attracts more research talent and enhances capacity in cooperative projects, competition for grants, raising donations and drawing good quality foreign doctoral students. Well-funded research infrastructure allows universities to deploy their best performing staff so as to concentrate areas of strength, and to secure intellectual leadership at both national and global levels. Research supplies both the practical know-how and symbolic capital that helps keeps the research university at the cutting edge.

Unmodified positional competition becomes more intense over time as demonstrated by the history of higher education in the United States. More and more bright students press for entry into the top institutions, the gateway narrows and 'failures' accumulate. Modest institutions are emptied of custom and dignity. The downside is not just an unequal distribution of social opportunities (Teese, 2000), but the isolation of many of the fruits of intellectual life in a handful of hard-to-enter institutions. Former Harvard President Derek Bok talks about 'powerful elites who are insulated from competition and able to set their own terms in a world increasingly unrestrained about inhibitions about greed' (Frank & Cook 1995, p. 5). The steeper the distance between elite universities and others, the more that society values elite universities and the less it sees of their benefits. Skyrocketing tuition prices, permitted by monopoly status and fed by wasteful competition for signs of prestige, reinforce the social closure. Scholarships for needy students are never on a scale sufficient to reassert equality of opportunity.[14]

[14] This is the logic of a winner-take-all market (Frank, 2001). In the USA the proportion of the top scoring students applying to the elite sector continues to grow, so they become more concentrated, segregated from the also-rans who are crowded out (Frank & Cook, 1995, p. 12). The American winner-take-all market spreads across the globe, like the markets in film/television or popular music. When high value becomes centralised and concentrated in a small number of products, price differentials lurch upwards. Elite universities are caught in a wasteful 'positional arms race' in

The Binary Field of Higher Education

In his analysis of *The field of cultural production* (1993, pp. 38-39) Bourdieu argues that the field is structured by a polar opposition. At one end there is the sub-field of restricted production. At the other end is the sub-field of mass production tending to commercial production. Each sub-field has a distinct principle of hierarchization. In the aristocratic elite institutions, which shape the high value products, the principle of hierarchization is cultural status, autonomous and specific to the field. In the mass or 'popular' institutions hierarchization is shaped by economic capital and market demand, and the institutions are heteronomous; though from time to time mass institutions renew themselves by adapting ideas from the elite sector. Between the two sub-fields lie a range of intermediate institutions which combine the two opposing principles in varying and unstable degrees. This polarity is readily seen in national higher education (Naidoo, this volume) where it is explanatory of system tension and differentiation.

Vertical segmentation of institutions is inevitable under conditions of positional competition. Likewise, it is inevitable that upward mobility will be constrained. At the top the production of positional goods combines competition with oligopoly and market closure. The laws of a conventional economic market cannot operate. Status, mediated by cultural distinctions, is dominant. In elite institutions, the more intense is consumer competition for entry, the less the university finds it necessary to court the consumer in the conventional manner by dropping prices or providing more and/or better services. As long as prestige is sustained the consumer will follow. For every student who becomes dissatisfied with the cursory feedback for their work or the star professors they never see, there are half a dozen other potential students waiting at the gate. Marketing, which is often under-stated, focuses on signs of venerability (gothic buildings, scholars); though in a consumer culture it also resonates with student-centeredness. (It is here that the sensitivity of elite institutions to their 'customers' is expressed). Once a university obtains elite status, where the competition is fierce but closed, and status itself helps to recycle student custom and research resources, the reproduction of the standing of the university requires no more than ordinary prudence.

---

which their costs are escalated by bargains with star professors and spending on visible signs of prestige (twenty-first century sandstone). 'When the stakes in such contests are high, each contestant may face irresistible pressures to make heavy investments that in the long term turn out to be mutually offsetting' (Frank, 1985, p. 136). Hirsch argues that in a status market 'the connection between individual and aggregate advance is broken' (Hirsch, 1976, pp. 7-8). Adam Smith's invisible hand fails to maximize the general welfare. On the contrary: arguably, market competition in social status, in and through higher education, not only fails to advance the general welfare, it reduces it, suggesting that the policy objective in higher education should be to augment investment in joint education infrastructure that is designed to benefit all: 'When social interaction of this kind [positional competition] is present, individual action is no longer a sure means of fulfilling individual choice; the preferred outcome may be obtainable only through collective action' (Hirsch, 1976, p. 5).

Table 2.1. Typical segmentation of competition in national higher education systems

Segment 1 Elite research universities	Self-reproducing, combining historical reputation, research performance, and student quality/ degree status. Driven by status attraction/ accumulation not revenues per se. Non-expansionary in size. Limitless ambitions for social status and power. Wealthy. Relatively closed especially at the peak.
Segment 2 Aspirant research universities	Struggling to live as Segment 1 but unable to break in. Tendency to brain drain of best students and researchers to Segment 1. May engage in selected commercial activities to generate revenues, but not so efficient in commercial terms. Resource scarcity. Semi-open
Segment 3 Teaching-focused (university or other)	Student volume- and revenue-driven. Some are private for-profit institutions, or public sector operations with a large commercial component, tending to expand. High resource scarcity. Hyper-marketing and shaving costs/quality under market pressure. Open

Regardless of whether the system is a high tuition high aid regime, or a free tuition regime, the positional hierarchy in higher education is remarkably stable over time, compared to market leadership in other industries. Australia's leading institutions are all pre-1958 foundations. In the USA they mostly date back to world war 1.

In the lower echelons the laws of competition are different. Positional power falls away to nothing and higher education is more like a conventional economic market. In the sub-field of mass higher education, institutions whether non-profit or for-profit struggle to fill their places and secure revenues. Once achieved success remains provisional and contestable. Such institutions do not have the resources to build a major research effort. Their core business is vocational learning and often (whether or not it is designated as such) basic literacy. At best, when public funding cuts, hyper-marketing and competitive efficiencies do not place severe limits on programs, teaching is of high quality and programs for social groups traditionally under-represented in higher education are successful. But mass institutions never receive full recognition for their best quality work. In a positional market everyone instinctively centres 'quality' on the high-prestige universities where research is conducted and high value positional goods are secured. High classroom quality in teaching-only institutions is over-determined by low social status.

Intermediate institutions, combining some scarce high value places and some low value access places, find it hard to move up because in a positional market the number of prestige producers is strictly limited. They become 'second choice' producers, or specialists. Newer research universities struggle hard to break into the upper echelon, imitating its programs and ethos, but mostly fail. Often they provide openings for young academics and innovative programs only to see people and programs 'brain drain' to the elite universities once their success is established. Table 2.1 summarises the segmentation of institutions in national systems. The top tier consists of high value positional competition marked by scarcity, exclusion and research intensity. At the bottom is high volume basic education, under-funded by governments, often produced in quasi-commercial or commercial markets, and marked by place-filling, expansionism and low unit positional value.

In the absence of egalitarian policy, over time scarcity and exclusion drives further vertical segmentation. Stratification is played out in the tense middle zone, in Table 2.1 the subordinated research universities. These are over-determined not only by insufficient and insecure status but by revenue hunger in a partly entrepreneurial setting. This steers research and scholarship away from long term inquiry to immediate and saleable applications (Marginson & Considine, 2000).[15]

Stratification is both formal and informal, varies by nation, and can be more complex than the Table suggests. Geiger (2004) cites seven segments in the USA. There can be horizontal variants, such as the differing paths to elite formation in U.S. research universities and liberal arts colleges. But everywhere, firm barriers retard upward movement between the segments, especially movement into the top. The conservative character of the national positional hierarchy is another sign that positional competition does not function as a conventional economic market.

The Formation of Global Markets

The cross-border mobility of people and ideas in higher education has a long history. Greek, Latin, Sanskrit and Chinese were global languages of scholarship long before English. Through much of the twentieth century there was significant migration of European academic labour into the U.S. universities; after world war two scholarships for students from emerging nations came to constitute an important component of foreign aid by English-speaking and European nations; and private student traffic also increased. By 1975 approximately 600,000 students were enrolled outside their country of citizenship (OECD, 2006, p. 286). But it is only in the last two decades or so, the period of accelerated globalization, that a clearly identifiable global market in higher education has emerged; and only since 2003 has the global market been rendered ultra-visible to all in the form of global university rankings (SJTUIHE, 2007; Times Higher, 2007; Chapter 3 this volume).

The formation of the global market has been driven by four factors. First, the expansion of cross-border mobility in the workforce; due to the increasing inter-penetration of economies by foreign capital, the growing number of products and production chains for which the world is a single market, the expanding middle classes in China, India, Korea and other Asian nations (which fosters demand for global products and enlarges the pool of potential global students) and the growing

[15] Nevertheless, the encounter between commercial and quasi-commercial markets and older professional practices is not pre-determined. In many institutions where the role of market competition is growing, semi-independent academic cultures, and restraint on growth so as to sustain prestige, are a brake on free-wheeling entrepreneurialism. Revenue is unambiguously supreme only in fully capitalist institutions. As Marx put it: 'Trade always has, to a greater or lesser degree, a solvent effect on the pre-existing organisations of production, which in all their various forms are principally oriented to use value. But how far it leads to the dissolution of the old mode of production depends first and foremost on the solidity and inner articulation of this mode of production itself' (Marx, 1981, p. 449).

economic scope for bi-cultural skills. These factors have increased the scope and value of portable qualifications, especially in business, technologies and scientific research. This has markedly expanded the scope for the production of global private goods in higher education. The dominance of English in technologies and business (not to mention consumption and entertainment) feeds into the demand for English-language degrees and immersion in English-speaking countries during the period of study. In developing and emerging nations, in which the quantity and/or quality of higher education is constrained and there is a lag before an education and research system of scale can be created, many governments see foreign education as a useful adjunct to national capacity (OECD, 2004a). Often foreign study is used to differentiate the career structures of government and national education system. Where offshore places significantly extend national capacity and individual choice in importing nations, this cross-border education constitutes not just private goods for the students who study abroad but global public goods in the nation; though high private costs may tend to reduce the distributional 'publicness'; and in nations where a foreign degree carries higher prestige than degrees obtained at home, the cross-border education of some might devalue the local private goods of others.

Second is the equally rapid growth of a commercial industry in cross-border degrees. British and Australian institutions have led the way here. The international agencies have expended much energy in policy discussions designed to facilitate cross-border trade, especially the World Trade Organization through the General Agreement of Trade in Services (WTO/GATS, 2005), the World Bank, and the OECD (2004a; 2004b). Global positional competition has been partly economised. Private investment has grown; and many producers expand places, minimise costs and maximise revenues and market share as in capitalist business; though commercialism remains a minority approach in the USA and Western Europe.

Third, and decisively, there is the formation of the worldwide information and communications system. Electronic networking has broadened the scope of both national and global status markets, enhanced the visibility of producers and centred and concentrated superior prestige on a small number of market leaders. As Frank and Cook (1995) note in relation to the American domestic market, the emergence of a single networked system on a larger scale increases the relative rewards to leading institutions and staff and reinforces the tendencies to market segmentation inherent in status competition. Demand for Ivy League institutions grows and these become more visible and less accessible. So it is in the global market.

The fourth factor is a technology conditioned by this world-wide communicative space: global university rankings, which as noted codify and legitimate the global market and normalize the global dimension as the horizon of comparison.

In the manner of all global phenomena the global market in higher education extends not to the totality of higher education but to the cross-border aspects. It is a market in mobility. Global positional goods are foreign degrees, accessed across borders and capable of use in more than one nation, including foreign doctoral training which takes place largely in the USA and underpinned by the most globally integrated aspect of higher education, research. The global market does not absorb the place-bound national markets in higher education, though it overshadows and affects them. Like them it is a market in positional goods. Its hierarchy of values derives from the hierarchy of institutions. This is shaped by four elements: the distribution of research capacity, which is even more important in the global market than the national markets because inherited tradition plays a lesser role; the hierarchy of nations, in which developed economies have a decisive advantage; the global dominance of English which has become the one common language in education and research as well as business and government; and above and beyond these factors the special and overwhelming hegemony exercised by research universities from the United States, particularly the private Ivy League, the University of California system and the other major state universities.

Structure of the global market. The polarity Bourdieu that identifies in national educational and cultural systems shows itself more clearly in the global field. At the elite end of the field is the 'Global Super-league' (The Economist, 2005): Harvard, Stanford, MIT, Yale, Princeton, Caltech, Chicago, Pennsylvania, Berkeley and other leading lights of the University of California system, the large mid-west universities and others in the United States, plus a handful in the UK led by Cambridge and Oxford. These universities, which constitute the sub-field of elite universities in their own nations, have concentrated knowledge power and hence status power not just on the national scale but on the world scale. With their 'brands' recognized everywhere they have become universal objects of desire and universal magnets for talent. At the opposite end of the global field are institutions focused on revenues, cost management and expansion in the manner of commercial companies. As noted, the sub-field includes not only the for-profit University of Phoenix and the global e-learning enterprises, but also many non-profit universities where they provide international education on a revenue-raising commercial basis.

In the intermediate zone between the sub-fields lie universities that vary in the extent to which they practise global research and status-building, and/or understand the global dimension simply as a field of business opportunities (Table 2.5 below).

The polar nature of the field of global higher education powerfully reinforces the global hegemony of United States' doctoral universities, predominantly located in the global sub-field of elite university education and research. The Bourdieuian theorisation of the field helps to explain why the American universities sustain a dominant global position without aggressively building their global operations as others do. They do not need to work hard to sustain the status of global elite, which rests ultimately on American economic, technological, cultural and military power and is reproduced as a matter of course as U.S. universities go about their daily domestic business. The Bourdieuian binarism also helps to explain why the global rankings systems have thrown up the techniques they have, including rankings of research performance and indicators of reputation and 'internationalisation'. Research rankings identify and service the elite sub-field. Reputational rankings construct a hierarchy of positional standing across the whole global field, taking in both the elite sub-field and the mass commercial providers. Quantity indicators of 'internationalization' service the mass market in foreign degrees (see chapter 3).

Nevertheless in global higher education the polarity between the two principles of hierarchization works somewhat differently to the market in art outlined by Bourdieu (1993). The elite sub-field in higher education is more robust than elite cultural producers, and more closely integrated with economic and political power. For the most part the super-league universities are economically stronger than the mass producers. Further, in higher education the tension between Bourdieu's two principles of hierarchization is absorbed not just between different types of university but also inside the elite universities, for example in research that is alternately fundamental and commercial in temper, as in bioscience (Bok, 2003); and in non core profit-making businesses untidily tacked onto core operations. At the same time Bourdieu is right to argue that the more autonomous that universities become, the less they are commercial in temper. The ultimate rationale of the global Super-league is not revenues but imperial prestige and knowledge power. The drivers are knowledge production, and the reproduction of social position. By sustaining themselves at the peak of the American domestic market the leading US universities can exercise global leadership without much additional effort. US hegemony in higher education is further explored in the final section of the chapter.

DATA CONCERNING GLOBAL MARKETS

The political economy of global educational markets is a novel area. Data in three areas can facilitate understanding: cross-border student movement into educational programs; short and long term movements by academic staff and doctoral students; and the differentiation of cross-border markets between institutions (and nations); including the distinction between the two sub-fields within the global market.

Cross-border Student Movement

There are many forms of student mobility and cross-border access, ranging from foreign experience acquired as an adjunct to home-based study in short stays of half a year or less; to 'twinning' programs whereby the student does part of the program in home country and the other part in a partner institution in the exporting country; to movement into the exporting country for the full length of the program, the principal form of cross-border education in terms of numbers; to cross-border education acquired from within the home country as distance education, or from a foreign campus on home soil operated either by the foreign institution or a partner organization. 'Branch campuses' play a growing role, with policy cooperation from the importing nation, in Singapore and Malaysia, and to a lesser extent in China. Cross-border distance education is mostly provided online, often with links to locally-based study centres. Global 'e-U's' have not grown as rapidly as many English-language universities and e-learning corporations expected, in part because ICT capacity is narrowly distributed outside in the developing world; in part because the export nations have yet to design curricula linguistically and culturally tailored to non English-speaking populations, in part because online degrees do not have equivalent status with face-to-face degrees. However this medium will become more important in future (Marginson, 2004; OECD, 2005a).[16]

Data concerning cross-border student movement are collected and published by national agencies and standardised by the OECD (2006). There are problems of category ambiguity and incompatibilities between national collections. Data do not always distinguish short-term from medium-term mobility. Free movement within the European area makes it impossible to use visas as a tracking device (OECD, 2006, p. 285). No country collects data on foreign distance education accessed by its citizens. A key difficulty is that only some countries collect data in relation to foreign students who cross into the country for educational purposes. Other nations distinguish citizen students from non-citizens but the latter may include students from families resident for a long time without citizenship, such as guest worker families in Germany (Kelo et al., 2006; p. 3). The OECD distinguishes 'international' border crossing students from 'foreign' students (OECD, 2006, p. 285). Another problem is that not all nations identify cross-border students who subsequently become migrants. Fortunately the principal target of educational migration, the USA, has an exemplary data base in this area (NSB, 2006).

The OECD data indicate that in 2004, 2.7 million students enrolled outside the country of citizenship, compared to 1.9 million in 2000. Of the 2.7 million students in 2004 just over half were from Asian nations, including 381,330 (14.4 per cent) from China and 129,627 (4.9 per cent) from India. A further almost one quarter of all cross-border student movement took place between European countries.

In higher education the rhetoric of 'internationalisation' in the exporting nations norms global engagement as a two way flow premised on mutual cultural respect. The reality is different. Global competition in degree programs takes the form of an export-import market in positional goods, characterised by uni-directional student flows and asymmetrical cultural transformations. Some nations are primarily exporters, others are primarily importers. A small group including Japan, Malaysia, and parts of Europe exhibits a more balanced two-way exchange. Table 2.2 lists the larger national higher systems that produce global positional goods (the exporters), and the nations of origin of the students that obtain those goods (the importers).

The data in the table are shaped by three factors. First, two way student movement in Europe, mostly consisting of short-term visits by students that will eventually return to their countries of origin. Intra-European traffic, enhanced by the ERASMUS program, is now underpinned by the momentum of Europeanisation. Second, the leading role of English language nations in exports. These countries command almost half of the world-wide total (46.8 per cent) between them. In 2004 the number one destination country was the United States with 572,509 students, 21.6 per cent of the total. Note also that France plays a special export role in relation to Francophone Africa, led by Morocco and Algeria (25,783).

---

16 For more discussion of the different modes of cross-border education, and issues of trade and mobility regulation in the importing and exporting nations, see among others OECD, 2004a; McBurnie & Ziguras, 2006; Sidhu, 2005.

Table 2.2. Principal export and import nations in the education of foreign students, 2004

Export nations	*incoming students: number of these students*	*proportion of foreign students %*	*Import nations*	*outgoing students: number of these students*	*proportion of foreign students %*
USA	572,509	21.6	China	381,330	14.4
UK	300,056	11.3	India	129,627	4.9
Germany	260,314	9.8	Korea	98,103	3.7
France	237,587	9.0	Germany	61,845	2.3
Australia	166,955	6.3	Japan	61,437	2.3
Canada	132,982	5.0	Morocco	58,038	2.2
Japan	117,903	4.4	France	57,231	2.2
Russian Fed.	75,786	2.9	Turkey	54,381	2.1
New Zealand	68,904	2.6	Greece	51,138	1.9
South Africa	49,979	1.9	USA	46,547	1.8
total	*2,651,144*	*100.0*	*total*	*2,651,144*	*100.0*

Source: OECD, 2006 and OECD data base.

Third, the importance of Asian source countries that constituted four of the five largest importers. Global competition in degree courses is largely centred on the movement of Asian students to the Anglophone zone and Western Europe. As well as China, India, Korea and Japan, the importers include Malaysia (42,054), Hong Kong China (36,186), Indonesia (33,877), Thailand (24,677), Singapore (21,163) and Vietnam (17,089). These students are mostly self-financed, with the notable exception of those at doctoral level. In much of Asia the habit of private investment is entrenched. In Korea 70 per cent of local expenditure on tertiary institutions is private, in Japan and Indonesia each 56 per cent, China 43 per cent (OECD, 2003).

In the Asia-Pacific region there is much potential for the further growth of education as a global positional good. These nations have well over half of the world's population and ten of the 16 cities with over ten million people, immense concentrations of present and future demand for education. Table 2.3 (over) lists the growing mega-cities. China has seen two decades of high economic growth and could produce one fifth of world GDP by 2050. Although domestic participation in higher education has tripled since the mid 1990s foreign education will continue to constitute positional value in China, especially in internationalised sectors such as finance and tourism, and open migration routes abroad. Parallel cases Korea and Japan have enough good quality tertiary places but positional demand for foreign education continues.[17] Further, unmet demand will continue to be a factor in some regions in China. Thailand and Indonesia are likely to see growing unmet demand.

[17] Once the acquisition of foreign education becomes a normal practice of middle class families in parts of mainland China and Southeast Asia, it becomes not so much a method of gaining a special

Table 2.3. Cities projected to have more than 10 million people in 2015

		2000 millions	*2015 millions*
Asia-Pacific countries:			
Mumbai	India	16.1	22.6
Calcutta	India	13.1	16.7
Delhi	India	12.4	20.9
Dhaka	Bangladesh	12.5	22.8
Karachi	Pakistan	10.0	16.2
Shanghai	China	12.9	13.6
Beijing	China	10.8	11.7
Tianjin	China	9.2	10.3
Tokyo	Japan	26.4	27.2
Osaka	Japan	11.0	11.0
Jakarta	Indonesia	11.0	17.3
Metro Manila	Philippines	9.9	12.6
other countries:			
New York	USA	16.7	17.9
Los Angeles	USA	13.2	14.5
Sao Paulo	Brazil	18.0	21.2
Rio de Janiero	Brazil	10.7	11.5
Mexico City	Mexico	18.1	20.4
Buenos Aires	Argentina	12.0	13.2
Lagos	Nigeria	8.7	16.0
Cairo	Egypt	9.4	11.5
Istanbul	Turkey	9.0	11.4

Source: ADB, 2003

European and Asian nations offer selected English-medium programs to complete for these Asian students, especially Denmark, Finland, the Netherlands, Sweden and Malaysia (OECD, 2006, p. 291). However, on the whole the English-speaking countries retain a global advantage vis a vis other export nations because they provide immersion in an English speaking setting on a daily basis, and offer significant opportunities to skilled migrants, encouraging foreign students to stay on after graduation (Trembaly, 2005). A clear-cut migration route substantially enhances the private value of cross-border education as a positional good. When citizenship in the country of education also opens the way to easier migration into other sought-after jurisdictions – for example entry to Australia via initial entry to New Zealand, entry to the USA via Australia or Canada – the positional goods constituted by cross-border education gain still further in value.

and usual advantage, as a 'defensive necessity' (Hirsch, 1976) for reproducing social position and sustaining the effectiveness of family business activities.

Dynamics of the market. Table 2.2 is limited as a guide to the market. It does not demarcate commercial cross-border education from subsidised forms.

Numbers are growing most rapidly where cross-border education is a commercial business with full-price tuition, in the UK, Australia and New Zealand. Malaysia (30,407 foreign students in 2004, mostly in the private sector), Singapore, China and India are emerging commercial exporters. In Australia, New Zealand and the UK, full fee foreign students are enrolled alongside subsidised domestic students in non-profit institutions, so that the commercial business is an island within the larger setting. Here the production of global private goods rests on the sunk capital and reputation of the public university, derived in large part from its long-standing and continuing function as a producer of public goods. It is another example of public/private symbiosis. However there can be tensions between the logic of the business, which tends to standardise mass market quality and empty out costs and complexity, and the requirements, rhythms and cost structures of bundled non-profit teaching and research. Further, if commercial full fee education becomes too dominant within the mix, to the extent of say a third or a half of overall student enrolments, the public good reputation of the institution begins to corrode. There is the danger of killing the goose that lays the golden eggs.

At the same time capitalist economic competition is not the only potential driver of growth in cross-border education. Europe shows this. The cultural momentum of Europeanisation (more solidly entrenched in education than in government and politics) sustains a common education policy framework that fosters large scale, growing and reciprocal flows of students and also staff. Likewise, though Japanese higher education is highly competitive and also privatised the Japanese government emphasises foreign education as a means to learning and to internationalization; that is, as a national and global public good not a revenue-generating business. The government subsidises student flows in both directions. No other government in Asia fosters student mobility on this scale and cooperative regional schemes remain modest or embryonic. In other cases non-commercial mobility is associated with modest growth rates. In the U.S. doctoral sector, France, German and much of Europe the great bulk of foreign students continue to access the exporting country on a non-commercial basis. In parts of Europe, such as the Netherlands and Denmark, the commercial component is increasing though provision is mostly non-commercial. The USA enhances the public goods component of cross-border education by treating it as an arm of strategic international policy and in part as foreign aid. Interestingly, it is unlikely that relatively generous access conditions provided in particular national systems such as Germany and the USA reduce the potential value of foreign education as a private good; because this is determined primarily by the relative prestige of institution and nation of education, not price.

Table 2.2 does not differentiate between elite provider institutions and providers of mass higher education to foreign students. However, the OECD data make it possible to identify or approximate the degree to which national systems are affected by cross-border flows, and also provide detail of the character of intakes as to level and field of study, both of which have some bearing on the question of elite versus mass provision. International students form a high proportion of total student enrolments in Australia (16.6 per cent), the UK (13.4 per cent), Switzerland (12.7 per cent) and Austria (11.3 per cent). Additionally, foreign student as a proportion of enrolments exceed 10 per cent in Canada, France, Germany and New Zealand (where the ratio is 28.3 per cent). In the largest exporter, the United States, international students constituted only 3.4 per cent of enrolments in 2004 and were relatively marginal in most institutions. The largest cross-border enrolment was in the University of Southern California which nevertheless had less than 7000 international students. On the other hand the American doctoral sector enrols a proportion of academically elite students which is exceptional by international standards. Nearly one third of foreign students are at doctoral level (IIE, 2006). This points to the overall global role of the United States as an elite provider (discussed further below); though some U.S. community colleges engage in volume building mass provision.

The cross-border mission of the U.S. doctoral universities contrasts with their Australian counterparts. In Australia international students were 16.6 per cent of enrolments and a low 3.9 per cent of those students entered advanced research programs, signs of a nation predominantly committed to mass production of global private goods as its principal international strategy, except at the flagship research university, the Australian National University (Marginson, 2007c). In 2005 in Australia seven universities enrolled more than 9000 international students, led by Monash University at 17,168. In Australia in 2005, 15 per cent of higher education revenues came from international students fees. The highest was 46 per cent of all income at one institution, Central Queensland University (DEST, 2007), where the identity of the original public sector domestic university has been partly displaced. In 2007 it faced bankruptcy after fee-paying numbers dropped by almost 3000. At that level of commercialisation, not only is the tail (the commercial business) wagging the dog (the public university), it has started up a second line of barking. In the UK there were 300, 056 international students in 2004, sign of a mission that is mass and commercial in character, but 11.5 per cent of international students were in advanced research programs in the UK. The UK provides for both Bourdieuian sub-fields of the global market (OECD, 2006, pp. 303 & 310).

Mobile Researchers

Data on academic staff movement. Luitjen-Lub et al. (2005, p. 157) remark that 'there are not many secure data on staff mobility'. For OECD Mahroum (2001, p. 220) notes that 'the proportion of overseas members of academic and scientific staff in national research institutions', that is, long-term academic migrants, is a useful indicator but 'is often neglected due to the lack of consistent, comprehensive and internationally comparative data'. However, there are better data on academic and researcher mobility in the areas of short-term visits and doctoral education.

The overall trend is one of increase in cross-border movement for research collaborations, conferences and short exchange visits; recruitment into academic posts – though the picture is uneven, systems ranging from relatively open, to closed to foreign entrants - and doctoral programs (OECD, 2004c; Enders & de

Weert, 2004; Marginson, 2008). Mobility has a long-standing place in the culture of research-intensive universities, though this varies by nation and by discipline. Research is often the medium for mobility because of the universal character of many knowledges especially the sciences. Teichler (2004) remarks that 'most academics hold cosmopolitan values in high esteem' and internationalizing one's knowledge base is a relatively 'safe' method of intellectual growth engaged selectively and at will (Teichler, 2004, p. 11). Avveduto's (2001) study of faculty and doctoral students in six Italian universities found that 'overseas experience is rated by the vast majority of professors and students as highly desirable and is often cited as a value per se'. Only 26 per cent of the students had studied abroad but 96 per cent of those who had not wanted to do so (Avveduto, 2001, p. 233). For many non-American researchers a period of international doctoral study or post-doctoral work, particularly in the USA., can both open up opportunities abroad and enhance the positional value of qualifications and vitae at home. This does not mean that the American or other foreign labour markets constitute a unitary global labour market that is subsuming the national academic career structures and labour markets. National labour markets remain robust and distinctive as to informal and formal rules, salaries and conditions of work, especially in larger national systems such as those of the UK, Japan, France and Germany (Musselin, 2004; 2005; Marginson, 2008). It is only in the market for highly paid high performing labour in scientific research, and in the market in doctoral education (see below), that specifically global markets have decisively subordinated national markets.

The USA, followed by the UK, draws the most visiting faculty. In the USA between 1994-1994 and 2004-2005 the number of international scholar visitors rose from 59,981 to 89,634 (49.0 per cent (IIE, 2006); though, similarly to the foreign student intake, it faltered temporarily after 11 September 2001. Within Europe the main receiving countries for researchers are the UK (30 per cent), France (15 per cent), Germany (13 per cent) and the Netherlands (10 per cent) (Luitjen-Lub et al., 2005, p. 157). Most European nations report growth in short-term faculty visits. One such case is Norway where 'there has been a substantial increase in all types of journey from 1981 to 2000', about 20 per cent (Smeby & Trondal, 2005, pp. 456-457). Different types of international travel are interrelated and mutually reinforcing (p. 456). In the 1990s visits related to cross-border research collaboration increased more rapidly than any other category (p. 457).

The global doctoral market. There has been a parallel increase in doctoral student mobility. Many governments subsidize foreign PhD experience, while 'universities that once largely recruited doctoral candidates locally are increasingly active on the national and the international market' (Enders and de Weert, 2004, p. 146).

On the provider side, world-wide doctoral education is dominated by the special global role of the United States. It is in doctoral education, where, ironically, cross-border student mobility is less commercial in character than in first degree and Masters education, that the first unitary global market has emerged. Whereas the cross-border market in first degrees and business Masters functions as an adjunct to the national markets of many exporting nations, being both marginal to those nations and to some extent fragmented between them, the cross-border market in doctoral degrees is both dominant vis a vis national markets outside the United States and a coherent, unified primary market in its own right. This arises from the concentration of much of the world capacity in basic research in the United States, and a deliberate U.S. policy of opening U.S. universities to foreign talent. The USA has over half of the top 100 research universities (see below) and provides a comparatively broad field of career and employment opportunities. In other words the American brand of doctoral education is notably superior as a positional good compared to doctoral studies elsewhere. The outcome is that the American doctoral market functions as the global doctoral market. Other national doctoral markets are adjuncts to this global (American) doctoral market. It is an imperial relationship.

In the USA in 2004-2005, 18.1 per cent of foreign students in higher education were at doctoral level, and 30.8 per cent in research-intensive universities. Whereas in 2003 the UK had 23,871 foreign doctoral students, Spain 11,765 and Sweden 3205, in 2004-2005 the American doctoral sector enrolled 102,084 foreign doctoral students. Almost three quarters received scholarships or other subsidies, mostly from their American universities (IIE, 2006). The foreign-born proportion among doctoral graduates rose from 13.5 to 28.3 per cent between 1977 and 1997. In mathematics and computing it rose from 20.2 to 43.9 per cent, in engineering 32.1 to 45.8 per cent (Guellec & Cervantes 2002, pp. 77-78). It is a positional 'win-win' for both parties. The American economy and research system is boosted by foreign talent, helping it to sustain global leadership. Foreign entrants have a range of life opportunities unavailable anywhere else. During their studies foreign students make a key contribution to American universities as research and/or graduate teaching assistants. Between 1985 and 1996 the number of foreign students primarily supported as research assistants rose from 2000 to 7600 (Guellec & Cervantes 2002, p. 89). Later the U.S. immigration regime places a Green Card within reach. Growth in the number of foreign doctoral students and their share of American PhDs has been matched by the growth in their propensity to stay. From 1987 to 2001 the stay rate for foreign doctoral graduates rose from 49 to 71 per cent (OECD 2004c, 159).[18] Stay rates are particularly high for doctoral graduates from China, India, Israel, Argentina, Peru, Eastern Europe and Iran; and also for some developed countries including the UK, Canada, New Zealand and Germany. In 2003 three quarters of EU citizens who obtained a US doctorate said they had no plans to return to Europe (Tremblay, 2005, p. 208). Not all foreign PhDs go on to work in higher education but between 1975 and 2001 there was a sharp rise in the foreign born with US doctoral degrees as a proportion of academic labour, from 12 to 21 per cent (NSB 2006, p. A5-45). American universities are more flexible and

[18] Stay rates are particularly high among the large number of graduates in engineering, computing and technologies (Gupta et al., 2003). While in 1985 50.0 per cent of foreign science and engineering doctoral degree holders planned to stay, by 1995 it had reached 70.6 per cent (OECD, 2002, p. 49).

open than the academic labour markets of most other nations. Maintaining this flow of talent is a high priority despite post-2001 hiccups in the visa process.[19]

At postdoctoral stage the USA offers the majority of posts worldwide. There is throughput from American doctoral study to post-doctoral positions, and a further group of people enter the U.S. for the first time after completing a doctoral degree elsewhere. Whereas recent studies in Europe suggest postdoctoral mobility is stable (Enders and de Weert 2004a, pp. 146-147) an increasing proportion of postdoctoral personnel holding US doctoral degrees are foreign born: between 1985 and 2001 this rose from 21 to 41 per cent (NSB, 2006, p. A5-47), indicating growing movement into the United States.

Other foreign doctoral graduates from American institutions return to the nation of origin or migrate elsewhere, most of them carrying a degree of commitment to American norms in higher education and research,[20] often taken into positions of governmental or institutional leadership in the home country. Thus American knowledge goods and models of higher education and research have continuous effects in most other national systems. However, the reverse is not the case.

Differentiation of the Map of Institutions

Like the positional value formed in national competition, global positional value combines degree/brand status with research performance/reputation. As noted, even more than at national level the means of differentiating institutions is research status. Strong research universities are strong attractors of cross-border students. Research status is primarily determined by the quantity and quality of publications and citations - which assists the relative position of English language speaking institutions because English is the sole global language of research publication – and research-related income and success in the struggle for competitive grants.

In 2003 the Shanghai Jiao Tong University Institute of Higher Education (SJTUIHE) issued the first authoritative world-wide survey of universities based on research performance. The SJTUIHE ranked institutions from 1-500. Place-by-place rankings were provided for the top 100 and the remainder were grouped in fifties and hundreds. These rankings have since been issued on an annual basis. In 2007 the SJTUIHE also issued discipline-based rankings in physical sciences, engineering, life sciences, medicine and social sciences. The SJTUIHE uses a composite index based on the university of education of winners of Nobel Prizes and Fields Medals in Mathematics, the university where the prize winners currently

---

19 After 2001 the recruitment of research students into the USA was more robust than the recruitment of other foreign students. In 2004-2005 the total foreign enrolment fell by 1.3 per cent but the number of doctoral students rose by 2.0 per cent (IIE, 2006).

20 The USA is also formative of the academic profession in other nations in another way. The 1992 Carnegie survey of the academic profession in fourteen nations identified the US as the main exporter of academic labour, supplying three of the nations surveyed – Hong Kong, Korea and Israel - with more than 18 per cent of their staff. The next largest exporter, the UK, is much less important. France and Germany also play a small role as exporters of academic labour (Welch, 2005, pp. 78-79). Most other nations are net importers of academic labour.

work, publications in discipline-based journals, publication in *Science* and *Nature*, citations, the number of 'HiCi' researchers (in the top 250-300 in the field) in the institution, and a final component that consists of these measures aggregated and expressed per head of academic staff. The measures favour large comprehensive research intensive universities that are very strong in science-based disciplines.[21]

Table 2.4. The world's top 50 research universities, based on measured performance, 2006

	institution	*pts*		*institution*	*pts*
1	Harvard USA	100.0	26	University Coll Lond UK	42.2
2	Cambridge UK	72.6	27	FIT Zurich Switzerland	41.2
3	Stanford USA	72.5	28	Washington, St. Lou USA	40.4
4	UC Berkeley USA	72.1	29	New York USA	38.4
5	Massachusetts IT USA	69.7	30	Rockefeller USA	38.3
6	California IT USA	66.0	31	Duke USA	38.2
7	Columbia USA	61.8	32	Minnesota, Twin Cit USA	37.8
8	Princeton USA	58.6	33	Northwestern USA	37.6
8	Chicago USA	58.6	34	Colorado-Boulder USA	36.4
10	Oxford UK	57.6	35	UC Santa Barbara USA	36.1
11	Yale USA	55.9	36	British Columbia Canada	35.5
12	Cornell USA	54.1	37	U Maryland Coll Pk USA	35.4
13	UC San Diego USA	50.5	38	Texas, SW Med C USA	35.2
14	UC Los Angeles USA	50.4	39	Texas, Austin USA	34.9
15	Pennsylvania USA	50.1	40	Utrecht Netherlands	33.4
16	Wisconsin/Madison USA	48.8	41	Vanderbilt USA	33.2
17	Washington, Seattle USA	48.5	42	Pennsylv State, U Pk USA	32.7
18	UC San Francisco USA	47.7	42	UC Davis USA	32.7
19	Tokyo Japan	46.7	44	UC Irvine USA	32.6
20	Johns Hopkins USA	46.6	45	Paris 06 France	32.4
21	Michigan–Ann A USA	44.5	46	Rutgers State, N Bru USA	32.3
22	Kyoto Japan	43.9	47	Southern California USA	32.0
23	Imperial College UK	43.4	48	Karolinska I Stoc Sweden	31.9
24	Toronto Canada	42.8	48	Pittsburg, Pittsburg USA	31.9
25	Illinois, Urbana-C. USA	42.5	50	Manchester UK	31.7

Source: SJTIHE, 2006

Of the leading 500 research universities in 2006, the USA had 167. Other major systems were the UK (43), Germany (40), Japan (32), Italy (23), Canada (22), France (21), China (19), Australia (16), the Netherlands (12) and Sweden (11). Only 21 of the top 500 universities are in nations where per capita GDP is below

[21] Of the 736 Nobel Prizes awarded up till January 2003, 670 (91 per cent) went to people from high-income countries, the majority to the USA, with 4 per cent from Russia/Soviet Union/ Eastern Europe and 5 per cent from emerging and developing nations (but note that nearly all the last group were working in the USA, UK or Europe when they did their Prize-wining work). People from emerging and developing nations have their best prospect of winning a Nobel Prize in Literature (10 per cent) or Peace (20 per cent) but these are excluded from the SJTUIHE index (Bloom, 2005).

the 2005 global average of $9420 (World Bank, 2006): 14 in mainland China, excluding Hong Kong, four in Brazil, two in India and one in Egypt.

The centralisation and concentration of research capacity is greatest at the top. All but seven of the top 100 research universities were in nations with per capita incomes of over $15,000 in 2005. The USA dominated with 54, including 17 of the top 20. The UK had 11 of the top 100, including Cambridge and Oxford in the top 20; the English speaking countries between them commanded 71 per cent. The relatively small national systems of Switzerland (four) and Sweden (three) are strong in the top 100 group. Canada, France and the Netherlands were in the top 50 (Table 2.4). Figure 2.1 shows the global strength of U.S. basic research.

Figure 2.1. The world's top 100 research universities, based on measured performance: distribution by nation

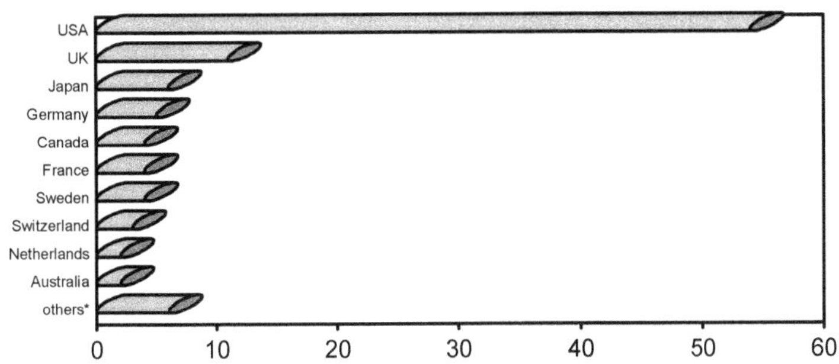

* Norway, Denmark, Finland, Italy, Israel, Russia each one.
Source: SJTUIHE, 2006

The leading research universities concentrate to themselves prestige, financial resources, human talent, research infrastructure and knowledge production; and each form of concentration tends to produce the others. Of the HiCi researchers in the top 250-300 in their fields, 3835 are located in the USA, almost nine times the number in any other country. The UK has 443, Japan 246, Germany 242, Canada 174, France 157, Australia 102, Switzerland 102 and the Netherlands 92. In the U.S. Harvard has 160 HiCi researchers, more than all French universities put together, Stanford 135 and UC Berkeley 82. There are 44 at Cambridge in the UK (Thomson-ISI, 2007). Where research capacity is concentrated, there knowledge flows are generated. In 2001 US scientists and social scientists published 200,870 papers in major journals. Indonesia with two thirds the U.S. population created 207 papers (NSB, 2006). Not much knowledge flows back from Indonesia to the USA.[22]

---

[22] The volume of papers from Japan was 57,420, the UK 47,660, Germany 43,623, France 31,317 and China 20,978 (NSB, 2006).

Yet research performance, while the strongest single element in comprising the global positional market, is not the only element. Above and beyond differentiation on the basis of research performance, institutions from the United States and to a lesser extent the UK enjoy a special global status. The universities from Canada, Japan and Western Europe in the top 50 are less known and less sought after. As noted, the global power of American universities is grounded not just in their intrinsic academic merits but rests also on broader American power, including its wide-ranging, flexible and well paid career opportunities. U.S. institutions also benefit from the manner in which they are projected through the world in American films and television. Is it possible for a European university outside the UK to be unequivocally ranked in the Global 'Super-league'? Probably, more so if it became Europe-wide in identity, but only with difficulty and after much preparatory work. Is it possible for a university from Japan or China to be so ranked? It would be harder, because of linguistic and cultural distances from the hegemonic case. One suspects the Universities of Tokyo or Peking could become world-wide household names only if the nation started to cut into the global role of the United States.

The other end of the global field. The mass and commercial sub-field is less readily identified than the 'Super-league'. The structural coherence of the market is limited. Cross-border education operates on the margins of most national systems; and far from being a common pool of activity it is fragmented between countries. What partly holds it together is that in the process of choice-making, students and their families can imagine a world-wide set of provider nations, institutions and places to live (an imagining powerfully assisted by global university rankings, not to mention the travel and tourism industries). Cross-border education is also partly unified by the practices of educational agents, who often work across several national jurisdictions and in relation to many institutions; and it is unified culturally by isomorphism in recruiting strategies and the provision of similar curricula in many countries in globalized fields such as technologies and business.

The mass and commercial grouping is a heterogeneous amalgam of institutions, with varied status at home within the domestic systems of the exporting countries but a more flattened status vis a vis each other in the trading environment. Precise gradations of position between institutions have less meaning than does national identity. Studies of student choice-making repeatedly confirm that below the top tier, students choose between one or another national system rather than identifying individual institutions (OECD, 2004a, p. 266); though if an institution is so fortunate as to figure well in university rankings it will use that in its marketing. In the mass and commercial sub-field, whereas on the demand side the driver is the desire for positional goods, on the supply side it is not institutional prestige but revenues. Between 1995 and 2003 average public funding per student fell 30 per cent in Australia and 12 per cent in the UK (Table 1.3 above). Between 1990 and 2005 the number of international students in Australia multiplied by 20. The commercial market expands in capitalist fashion because below the globally elite universities, an expansion in the number of places does not affect the unit value of each place.

Global segmentation. Table 2.5 summarises the segmentation of higher education when the global and national dimensions are both included.

Segment 1 is the elite sub-field of the global market, the 'Super-league'. The mass and commercial sub-field of the global market is split between segments 2 and 3. Institutions in segment 2 often experience internal tensions as to their cultures and missions. One example is those British and Australian universities competing in the global research stakes while at the same time building high volume concentrations of full fee-paying international students to plug the hole left by reductions in the government funding. When universities with high domestic status operate also in the mass international market they exhibit a Jekyll and Hyde personality, maintaining scarcity and hence positional value in the domestic intake, while expanding cross-border students in the manner of a business.

Table 2.5. Segmentation of global competition in higher education

Segment 1 World market of elite universities	The American doctoral sector and the high prestige universities in UK. Prestige not profit-driven. Prestige rests on research reputation and global power of degree.
Segment 2 Exporting national research universities	Research universities in the UK, Canada, Australia, Europe, Japan, etc. Prestige-driven at national level but may provide foreign degrees as a profit-making business.
Segment 3 Teaching-focused export institutions	Lesser status institutions in the export nations, operating commercially in the global market, catering to a lower cost/ lower quality echelon of foreign education.
Segment 4 Nationally-bound elite universities	National prestige providers, some research intensive. Little cross-border role. Nationally competitive with Segment 2; status vis a vis Segment 3 is a case-by-case matter.
Segment 5 Lesser status national/ local institutions	Confined to national competition and local demand. No cross-border role. Much the largest group of institutions, especially in the importing nations.

They attempt to play both sub-fields at once. But there are limitations to capitalist growth: the need to sustain a homogenous educational environment to protect domestic prestige (for example relatively high English standards for cross-border entry); and to maintain primacy of the research mission. If the commercial operation absorbs too much in development spending (for example in offshore sites) research performance will fall away. Teaching-only institutions can be fully commercialised but basic research producers cannot. Beneath that group in global status are ostensibly teaching-research universities for whom a research mission is useful in securing their marketing position but is is decisively subordinated to chasing cross-border revenues. Other national research universities located in segment 4, especially those located outside the English speaking world with less commercial opportunities in cross-border education, take the opposite tack. They continue to operate as elite universities but fall below the Super-league in research and have no presence in the global market for students.

Though the mass and commercial producers in Segments 2 and 3 may have lesser status than the Harvards and Oxfords, in the developing and emerging country contexts these institutions mostly enjoy a high status that is instrumental to the workings of the market. For foreign graduates returning to Thailand or Tajikistan, all reputable foreign degrees have positional value. All education export nations enjoy a generic positional advantage on the global scale. They benefit from the absence or weakness of higher education capacity in emerging nations, and the deflation of the value of higher education places in those nations by global referencing and foreign intervention. Without these factors students would have less opportunity to secure positional advantages through cross-border mobility. There would be less scope for the production of cross-border commodities, and a much smaller global economic market in mass cross-border higher education.[23]

NATIONAL DIMENSION AND GLOBAL DIMENSION

National and Global Markets

Global market and national markets are heterogeneous to but closely implicated in each other. The global market is not grounded in a global nation-state, or a global consciousness. Nor (with the marginal exception of the e-U's) does it offer global degrees. It offers the national degrees of one or another export nation. A cross-border degree is an American degree, an Australian degree, a German degree, a degree from Singapore. Especially in the USA, the main business of global provider institutions is not global positional competition but national positional competition. The national status they carry, and the status they confer on graduates in national markets, underpin their global value. But this is not to say that the global market is a weak epiphenomenon of the national market and still less that the global dimension is reducible to the national dimension. In all nations apart from the United States, the global hierarchy washes back into the national dimension and transforms and subordinates the national positional market.

There are two kinds of wash-back effect from the global dimension at national level. The more minor one is the impact of cross-border student populations on status, revenues, programs and the academic and organizational cultures. These students have differing implications for institutions compared to local students. In the national market, domestic students who are socially and academically selected augment the positional status of the institution. Attracting such students is a

---

[23] The greater the extent of global inequality and educational backwardness, the greater the gains made in border crossing and the more the potential for export profits[23]. As Marx notes: Capital invested in foreign trade can yield a higher rate of profit ... because it competes with commodities produced by other countries with less well developed production facilities, so that the more advanced country sells its goods above their value... (Marx, 1981, pp. 344-345).

principal strategic goal of middle and upper status institutions. Cross-border students have mostly been socially selected; the average cost of tuition and living expenses in the U.S. state university is six times the per capita in come of China (IDP, 2001); and they are often academically selected, particularly those that enter American universities. But with the exception of good doctoral students, cross-border students do not create status benefits additional to the revenues they generate.[24] The societies in which their families enjoy power and status are unknown to the domestic systems in which their cross-border education takes place. Even though it has become de rigueur for elite universities to be globally engaged, they are often surprisingly indifferent to the cross-border students in their midst. In the English speaking countries the flood of cross-border students on the whole has failed to trigger transformations in teaching and learning, for example culturally hybrid pedagogies and curricula. In Western Europe, the main effect has been to hasten the introduction of more English-medium programs. On the other hand cross-border students are often seen as a significant source of revenues.

The more important wash-back effect is the transformation of national positional markets. As noted, inside the USA global competition in higher education has little effect as such. Global competition is American competition and vice versa. But outside the USA there is a new global map of opportunity and this has remade the strategic terrain on which each national competition and each institution operates.

The global market does not replace the national markets. Rather, it subordinates them (again, except in the USA). Though the national and global hierarchies are imperfectly integrated they constitute a single set of possibilities for a growing number of students and families (they were long seen this way by academic staff and doctoral students). As Table 2.5 shows the effect is to layer on top of each national positional market a stratum of student places with superior positional value to *all* places created at the national level. The fact that few students from the nation will access cross-border places and much fewer will reach the Super-league does not matter. A market is an imagined field, and the global layer is now readily imagined. Few people know much about higher education in other nations but the peaks of global status are visible to all. Above the national hierarchy in every nation now looms the American doctoral sector and the leading UK universities. Global engagement varies from university to university but the potency of global referencing does not. In other words, once the point of reference is global rather than national, a university's national position is 'over-determined' by the global position occupied by its nation within the global higher education environment.

Paradoxically, the institutions that have most to lose are those historically strongest at the national level: those that educate the social elite, and the research intensive universities (often but not always the same institutions). The institutions that service the social elite, once unchallengeable positional leaders in their own domain, suddenly look less attractive and more vulnerable. The nation is no longer

[24] There are other exceptions to this generalization, in those elite institutions or units that specialise in educating foreign students, such as the London School of Economics, and to a lesser extent the Kennedy School of Government at Harvard.

solely *their* domain. They cannot compete with the charms of Oxford or Stanford. There are sudden leaks in the circular reproduction of status. Some of their treasured clients are crossing borders and slipping from their grasp. For research universities the problem is if anything worse, because it is almost impossible to compete with the American sector. Metrics and rankings are pitilessly transparent. Suddenly, local research performance becomes less worthy, and more specifically determining of status. This affects leading universities in nations such as Western Europe or Australia, which at least have the potential to become global players in their own right; but it is worse for the leading universities in emerging nations with less geo-strategic power, where even the elite institutions lack the capacity in research and communications technologies necessary to be globally effective.

This weakening of leading national institutions creates both 'public bads' and 'private bads'. On the public side it deforms national capacity in higher education. On the private side it subordinates at the global level the local elites left behind in the national institutions. Nevertheless, the 'wash-back' effects of global positional competition on national competition and national policy are no more equal or uniform than is the global market that generates them. Some local traditions are more robust than others, some nations and some institutions are more open to global influences than are others; and the capacity to pursue a proactive global strategy is unevenly distributed. Here national governmental strategies have a potential to affect the outcome. Some governments underpin the forward strategies of their leading institutions in the global setting (the wisest policy). Some offer domestic protection from those same global market forces.[25] Some unwisely leave it to the market to sort it out. Leading institutions still hold more cards than the others, even in developing nations. They can shore themselves up to some extent by working the national political system. The difference is that they no longer set the horizon of possibility. The ultimate constraints are now set by global segmentation.

So far this suggests that self-determination has been reduced. On the other hand global competition offers all institutions whatever their domestic status a new set of strategic options, identities and development paths. They can create international research partnerships (if they have research capacity), double-badged degrees with foreign providers, ICT-based linkages, revenue raising cross-border programs, set up off-shore and/or develop a more cosmopolitan curriculum. Operating in more than one national sphere they can use the outcomes of strategies in one sphere – resources, networks, reputation – as inputs in the other. (They also face new

[25] In many countries, protecting the strength of leading institutions is a key objective of policy. Those institutions can be shored up by intensifying their national positional value via competitive entry, or strong government investment in the science-based disciplines: both of these factors apply in the national universities in Japan. In Malaysia, despite the high reliance on foreign tertiary education, at undergraduate level the nationally dominant bumiputra are channelled into the leading public universities. This maintains the prestige of those institutions. The Chinese and Indian families make heavier use of undergraduate foreign education. The bumiputra benefit from government-funded scholarships for foreign education at postgraduate stage, offered to university academics and public servants marked out as future leaders.

tensions between domestic investment and global investment options). Some non-elite universities that are locked out of the elite segment of their national systems position themselves as providers of high value positional goods for students from elsewhere. Others build a new domestic status on the back of internationalized operations. At the same time, the national zone continues to be partly determining of global fortunes. Revenues continue to be sourced largely from national and local sources. Even Australia derived only 15 per cent of university revenues from teaching foreign students in 2005 (DEST, 2007). Offices and campuses established off-shore tend to lose money. Research universities spend more on globally-linked research activity than they generate in cross-border research funding.

The new salience of the global dimension also feeds into a changing distributional politics of higher education in many countries (OECD 2004a, pp. 241-248). It problematizes traditional national conceptions of equity policy. It is difficult to monitor equality/inequality of opportunity among the citizens of national education systems when there are leakages to foreign providers abroad and growing numbers of non-citizens at home. How can equity be established between domestic students attempting to access scarce places in national universities (for example professional training in Medicine or Law), and students who 'queue jump' by accessing such programs offshore, and cross-border students who enter the programs by different routes to domestic students? There are also questions about the social, cultural and gender composition of mobile student populations. What is the role of parental income and social status in determining effective access to sponsored schemes, and to full-cost foreign places? What are the longer term effects of differentiating access to foreign education among students in emerging nations, in the formation of societies in those nations?

The Hegemony of the United States

Globalization in higher education is what America does to the world not what the world does to America. It is the virtue of global hegemony (if virtue it is). American hegemony is exercised without entrepreneurial marketing. United States universities do not have to adjust their programs or their cultural ambience to attract international support. Few offer even a partly internationalised curriculum. Rather, they offer themselves as the global standard. Cross-border flows into the United States would be much greater if the U.S. doctoral sector was a more vigorous recruiter at first degree and Masters level. The overwhelming majority of prospective international students exhibit first preference demand for U.S. institutions, especially for the elite institutions (for example Mazzarol et al., 2001).

But the USA is concerned less with numbers than with the quality personnel needed to reproduce its global role. This already rests on a massive material base. The USA has the world's third largest population; its GDP is largest and GDP per head exceeds $40,000. Japan has less than half the population, one third the GDP and a per capita income of just over $30,000. The USA spends more of GDP on tertiary education than any other nation, 2.9 per cent, amounting to about $360 billion in 2005 (OECD, 2006). Japan spent $51 billion. In other words, the United

States invests more than *seven times* the whole of the rest of the OECD region. This is on par with the American global supremacy in military weapons and in film and television. It is not surprising the US has most of the top research universities. Table 2.6 compares the US to the next ten OECD countries and other large nations.

United States universities also dominate the institution-to-institution networking which structures the communicative field of global competition and enables it to be imagined. Universities in the different world regions normally have partial linkages with other regions but are always linked to universities in the United States, which is the global communications and business hub (Castells, 2001). Yet those same U.S. universities scarcely imagine the global field. With the national market also the top echelon of the global market they do not have to look further. There is intense domestic competition for top students, academic staff and research reputation but global competition does not generate a parallel vigour. Non-American universities carry little weight, and the dynamics of globalization, including the effects of U.S. universities abroad, are largely invisible.

Table 2.6. USA indicators for population, tertiary education and research compared to next ten nations in investment in tertiary education and five largest other nations in population, 2003/ 2007

	population 2005	GDP in PPP terms 2005	GDP per capita PPP 2005	investment in tertiary education 2003	investment in tertiary education# 2003/2005	Thomson-ISI 'HiCi' researchers 2007	number of research universities in Shanghai Jiao Tong University Institute of Higher Education ranking 2006		
	millions	billion $USD	$USD	% GDP	billion $USD		top 50	top 200	top 500
USA	296.5	12,409.5	41,854	2.9	360	3837	37	84	167
Japan	128.0	3943.8	30,811	1.3	51	246	2	9	32
Korea	48.3	1056.1	21,868	2.6	27	3	0	1	9
Germany	82.5	2417.5	29,309	1.1	27	243	0	15	40
France	60.7	1829.6	30,120	1.4	26	157	1	6	21
Canada	32.3	1061.2	32,885	2.4	25	175	2	8	22
UK	60.2	1926.8	32,007	1.1	21	444	5	23	43
Italy	57.5	1667.8	29,019	0.9	15	75	0	6	23
Mexico	103.1	1052.4	10,209	1.3	14	14	0	1	1
Spain	43.4	1133.5	26,125	1.2	14	18	0	1	9
Australia	20.3	643.0	31,642	1.5	10	105	0	6	16
China *	1311.4	8787.2	6701	n.a.	n.a.	18	0	2	14
India **	1095.6	3815.6	3483	0.7	27	11	0	0	2
Brazil	186.4	1627.3	8730	0.8	13	4	0	1	4
Russia	143.2	1559.9	10,897	0.7	11	5	0	1	2
Indonesia **	221.6	847.4	3842	0.3	3	0	0	0	0

PPP = Purchasing Power Parity. Investment in tertiary educaiton in dollars estimated using 2005 GDP data and the 2003 proportion of GDP allocated to tertiary education * includes Hong Kong, excludes five universities from Taiwan ** proportion of GDP spent on tertiary education is 2002 for India and Indonesia n.a. = data not available

sources: World Bank (2007), OECD (2006), Thomson-ISI (2007), SJTUIHE (2007)

American 'exceptionalism' in higher education is constituted not only by material weight but also by indifference. In scholarship, others cite Americans. Americans agree. The USA produced less than a third of the world's scientific articles in 2001 but 'accounted for 44 per cent of citations in the world scientific literature' (Vincent-Lancrin, 2006, p. 16). The Carnegie cross-country survey of the academic profession found that while over 90 per cent of scholars from other nations saw it necessary to read foreign books and journals only 62 per cent of American scholars agree. Despite the fact that many American scholars have exemplary engagements with foreign work this was much the smallest level among developed nations. Altbach remarks that though U.S. scholars are 'at the centre of the world academic system', 'the American research system is remarkably insular, especially when compared to scientific communities in other countries... The American system accepts scholars and scientists from abroad, but only if they conform to American academic and scientific norms' (Altbach, 2005, pp. 148-150).

Potentials for Pluralisation of Capacity

It is unlikely to stay this way for ever. The global hierarchy in higher education is subject to continual movement and flux. It is more unstable, more changeable, than national positional hierarchies in the industrialised nations; though as in the case of national competition the elite segment is more stable than others. The development of capacity in the emerging nations, especially research capacity, has the potential to modify the pattern of global asymmetries and uni-directional transformation.

European collaboration could alter the global picture in the longer term, given the research strength of several West European nations. Though it remains to be seen how those separated strengths will coordinate and accumulate, the Bologna declaration and the European Research Area provide favourable conditions for the concentration of capacity (Marginson & van der Wende, 2007). The more striking development is the rise of new science powers in East and Southeast Asia. Between 1988 and 2001 the number of scientific papers increased sharply in Korea (1332 per cent), Singapore (535 per cent), Taiwan (472 per cent) and China (354 per cent) (NSB, 2006). Singapore shows that an emerging nation can not only reverse the brain drain but can transform the global role of the nation by investing in globally focused education and research. The National University of Singapore might become one of the world's leading universities. China is rapidly creating a layer of top research universities. Research capacity in Hong Kong and Taiwan is at European levels. The long start of U.S. research universities, their control of the material means (research infrastructure, electronic publishing, journal production), and their capacity to co-opt foreign talent via hirings and collaboration, ensures their leadership of global knowledge flows for the foreseeable future. However, it is likely that an increasing number of researchers and scholars will move back and forth between systems during their careers and many will hold joint appointments; so that increasingly the American doctoral and post-doctoral experience will feed into capacity building back home, and more nations will build research systems capable of self-reproduction. Greater mobility and pluralisation of research capacity opens the way to a partial pluralisation of global status and in the longer term, the evolution of a more plural set of models and norms of institution in a more cosmopolitan higher education landscape (Marginson & Ordorika, 2007).

The key is the role of public investment by the nation-state. Public investment is the driver of change in China and Singapore and the mainstay of European research strength. Though the dynamics of global competition can only be modified by policies coordinated across borders; that is, global public goods; all nations have partial autonomy and scope for self-determination in the global setting (though the larger and wealthier nations, and universities, have more independent agency freedom than others). The effects of global competition in national systems in part can be modified by national government actions. By investing in global competence and strategies, nations and institutions are able to expand the range of possibilities. In essence, the more that global and national developments are driven by endogenous market competition in the sector, the more the outcomes of that competition will intensify the prior hierarchy of institutions and nations; and the less likely that new concentrations of research capacity and educational status will appear. Targeted public investment is the circuit breaker.

In general, developed nations have a superior capacity to access both global private and global public goods in higher education. They contain more people with the ability to pay for global private goods as foreign degrees or commercial intellectual property. They also have research infrastructures and trained personnel able to turn the public goods created by research into both more basic research, and technology transfer. In contrast, less developed nations benefit more from global public goods than global private goods. Higher education and research are integral to nation-building and to modernising strategies able to secure purchase in the global setting. Imports can make their optimum contribution to national capacity building when domestic infrastructure is already strong, the national innovation system is a magnet for diasporic investment, and the nation maximises 'brain return', as in Korea and Taiwan, making better use of its foreign-trained nationals. However in less developed nations, the cross-border education of nationals in the market for private positional goods is associated with brain drain, while PhD graduates who return often lack opportunities to work in their area of training. Cross-border education is less valuable to those nations than growth in higher education capacity at home. This, more than foreign education augments the pool of professional skills and the capacity of national research and knowledge systems, creating multiple long-term potential for national private and public goods.

Public investment is most vital in nations that are currently positioned altogether outside the research university circuit. Higher education should be provided as public goods in situations of market failure; and/or to increase the elements of non-rivalry and non-excludability in production and distribution; and/or to evade the opportunity costs and direct costs of marketing and competition (Singh, 2001).[26]

[26] Pusser notes: 'The fundamental arguments for public supply [i.e. non-market production by government agencies] are that it offers the most direct utilisation of public subsidies, and that it is the organisational type best suited to the rapid expansion of higher education… there is no diversion of the public subsidy to profit, hence more of the subsidy goes to the production of preferred goods' (Pusser, 2002). The argument is stronger if 'preferred goods' are externalities or collective goods.

Experience suggests that in the poorer nations the mechanisms of competition and markets will not deliver (Taskforce, 2000; OECD, 2004b). There is insufficient domestic private capital and its distribution is uneven. Only a small fraction of the population can access higher education via the private capacity to pay and such families typically move offshore for university education if not before, failing to contribute to a common national infrastructure. In these nations the way forward is to produce higher education as a predominantly public good, financed by state funding and aid monies under national/local control. This enables higher education and research to be mobilised for state-building and nation-building projects, while fostering spaces in which autonomous intellectual cultures can emerge.

The growth of the commercial cross-border market, in the context of neo-liberal policies, has been associated with a decline in foreign aid for higher education, especially aid from export nations such as the UK and Australia (OECD 2004a, p. 284-286). Developing nations have been rendered less aid-dependant and more market-dependant. Arguably, this has been exactly the wrong way to go.

By way of conclusion: Global public goods. As noted global public goods include both collective goods and positive global externalities (negative global externalities are 'public bads'). In extensively and intensively networked sectors such as higher education and research potentials for global public goods and 'bads' are enhanced. The most important global public goods are knowledge in the different fields, incorporating the potential for diverse collaborative conversations. At best doctoral universities function as highly cosmopolitan communities with many potential spin-offs for both the nation of education and for all nations. Higher education is a fecund site for association, linking not just members of kinship and affinity groups but erstwhile strangers. To borrow a term from the social capital literature, it creates 'bridging' relationships (Woolcock, 2001) across traditional social and cultural divisions. However higher education does not always function at its best. When annexed to hegemonic relations of power the leading research universities tend to impose a linguistic and intellectual mono-culture, crowding out research and scholarship in the other, non dominant languages and traditions. For subaltern universities it is empowering to connect more effectively to the global mainstream but de-powering if the price is a reduction in self-determination.

A good example, noted above, is the spread of English as a research language. Having a common language favours the rapid transmission and sharing of knowledge; but to the extent that it displaces knowledge created in other languages, its effects outside the English-speaking world are mixed. Global public goods can be exclusive as well as inclusive. Another public good flowing freely in the global setting, and mobilised for projects of cross-border control and transformation, is the Americanized model of system and institutional organization. When enfolded into the policy templates of global lending agencies, these models can undermine viable local traditions in education (Bensimon & Ordorika, 2006; Marginson & Ordorika, 2007). In other words, what constitutes a national and global public good from the viewpoint of the dominant nation or nations, might constitute public 'bads' from the subaltern viewpoint. The question to be asked is 'whose public?'

and 'whose global?' Another example is the cross-border mobility of people. This is a global collective good (in that it augments the common regime of choices and the stimulus of diversity), and a public good in those countries that benefit in net terms from the flow of talent. It is a 'public bad' in countries where the brain drain is increased. Global relationships are very different depending on whether they embody a singular national/imperial configuration of power, or a multilateral configuration of power. In a multilateral framework the potential for 'win win' global public goods is maximised. This points to the potential importance of multilateral agencies as embryonic global policy sites and to the need to ensure that these agencies are genuinely independent of individual nations and constitute a common space. This can help to overcome the principal difficulty in augmenting global public goods, which is the relative invisibility of those goods.

Here there are untapped potentials in the world-wide higher education setting. The Bourdieuian global market does not contain or exhaust all of the possibilities. By way of concluding remarks the chapter now moves from its primary focus on 'what is' in world-wide higher education to a consideration of 'what might be'.

Despite some discussion of issues such as recognition regimes and brain drain, global public goods in higher education are scarcely visible or understood; being under-recognised compared with national public goods, and global private goods generated in multilateral trade. The starting problem is that public goods can only be effectively considered and regulated in a policy space, and there is no such space ready-made for higher education. Higher education sits in a world that is increasingly inter-dependent yet defined by a zero-sum legal and geographical alignment, a world of autarkic and contesting states with no intrinsic incentive to cooperate despite accumulating problems with downstream effects (global climate change, ecological sustainability, water, population, urban crowding, rural poverty, epidemic diseases). There is a 'jurisdictional gap', a 'discrepancy between a globalized world and national, separate units of policy-making' (Kaul et al., 1999, p. xxvi). This again points to the need for multilateral processes, in two respects.

First, 'governments must assume full responsibility for the cross-border effects that their citizens generate' (Kaul et al., 1999, p. xxvii). This suggests units within each national government that take responsibility for positive and negative global externalities, and negotiate cost sharing with other governments. Second, there is a need for an inter-governmental space focused on higher education where the costs and benefits of global externalities are defined and managed, national governments are encouraged to incorporate cross-border externalities and prices into their routine national decision-making; and collective goods to be negotiated and developed, for example means of lowering barriers to global mobility. Such a global policy space might also consider issues of balanced global development, including the building of national educational capacity in the developing world, and the augmentation of cultural diversity in educational and linguistic contents. Such a process can be seen in embryo in UNESCO's discussions of quality assurance and in the OECD (2005) project on internationalisation in higher education. In Europe collaborative policy is more advanced. It is a sign of what collective action can achieve in higher education and research, if the will is there.

REFERENCES

Altbach, P. (2005). Academic challenges: the American professoriate in comparative perspective In: A. Welch (Ed.) *The professoriate: Portrait of a profession*, 147-165. Dordrecht: Springer.

Appadurai, A. (1996). *Modernity at large: Cultural dimensions of globalization*. Minneapolis : University of Minnesota Press.

Avveduto, S. (2001). International mobility of PhDs. In OECD, *Innovative people: Mobility of skilled personnel in national innovation systems*, pp. 243-260. Paris: OECD.

Beerkens, H. J. J. G. (2004). *Global opportunities and institutional embeddedness: Higher education consortia in Europe and Southeast Asia*. Doctoral; thesis. Center for Higher Education Policy Studies, University of Twente. Accessed 10 February 2006 at: http://www.utwente.nl/cheps/documenten/thesisbeerkens.pdf

Bensimon, E. and Ordorika, I. (2006). Mexico's estimulos: Faculty compensation based on piecework. In R. Rhoads and C. Torres (Eds.) *The university, state and market: The political economy of globalization*, 250-274. Stanford: Stanford University Press, Stanford.

Bloom, D. (2005). Raising the pressure: Globalization and the need for higher education reform. In G. Jones, P. McCarney, & M. Skolnik (Eds.) *Creating knowledge, strengthening nations: The changing role of higher education*, 21-41. Toronto: University of Toronto Press.

Bok, D. (2003). *Universities in the market-place: The commercialization of higher education*. Princeton: Princeton University Press.

Bourdieu P. (1984). Distinction: A social critique of the judgment of taste. R. Nice (transl.). London: Routledge & Kegan Paul.

Bourdieu, P. (1988). *Homo Academicus*. P. Collier (transl.). Cambridge: Polity.

Bourdieu, P. (1993). *The field of cultural production*. R. Johnson (Ed.). New York: Columbia University Press.

Bourdieu, P. (1996). *The state nobility: Elite schools in the field of power*. Translated by L. Clough. Oxford: Polity.

Breneman, D., Pusser, B. & Turner, S. (2007). *Earnings from learning: The rise of for-profit universities*. New York: SUNY Press.

Castells, M. (2001). *The Internet galaxy: Reflections on the Internet, business and society*. Oxford: Oxford University Press.

Calhoun, C. (1998) The public good as a social and cultural project. In W. Powell & E. Clemens (Eds.) *Private action and the public good*, 20-35. New Haven: Yale.

Deem, R. (2001). Globalisation, new managerialism, academic capitalism and entrepreneurialism in universities: is the local dimension still important? *Comparative Education*, 37(1), 7-20.

Department of Education, Science and Training, DEST, Australian government (2007). *Selected higher education statistics*. Accessed 17 June 2007 at: http://www.dest.gov.au/sectors/ higher_ education/ publications_resources/statistics/default.htm

Dow, S. (1990). Beyond dualism. *Cambridge Journal of Economics*, 14, 143-157.

The Economist (2005). The brains business, 8 September.

Enders, J. & de Weert, E. (2004a). Science, training and career: Changing modes of knowledge production and labour markets. *Higher Education Policy*, 17, 135-152.

Frank, R. (1985). *Choosing the right pond: Human behaviour and the quest for status*. New York: Oxford University Press.

Frank, R. (2001). Higher education: The ultimate winner-take-all market? In M. Devlin & J. Meyerson (Eds.) *Forum futures: Exploring the future of higher education*. Forum Strategy Series Vol. 3. Jossey-Bass: San Francisco.

Frank, R. & Cook, P. (1995). *The winner-take-all society: How more and more Americans compete for ever fewer and bigger prizes, encouraging economic waste, income inequality, and an impoverished cultural life*. New York: The Free Press.

Geiger, R. (2004). Market coordination in United States higher education. In P. Teixeira, B. Jongbloed, D. Dill, & A. Amaral (Eds.) *Markets in higher education: Rhetoric or reality?* Dordrecht: Kluwer.

Guellec, D. & Cervantes, M. (2002). International mobility of highly skilled workers: From statistical analysis to policy formulation. In OECD, *International mobility of the highly skilled*, 71-98. Paris: OECD.

Gupta, D., Nerad, M., & Cerny, J. (2003). International PhDs: Exploring the decision to stay or return. *International Higher Education*, 31(8). Accessed on 10 February 2006 at: http://www.bc.edu/bc_org/avp/soe/cihe/newsletter/News31/text008.htm

Harvey, D. (1990). *The condition of postmodernity : An enquiry into the origins of cultural change*. Oxford: Blackwell.

Harvey, D. (2006). *Spaces of global capitalism: Towards a theory of uneven geographical development*. London: Verso.

Held, D. & McGrew, A. (2000). *The global transformations reader*. Cambridge: Polity.

Held, D. McGrew, A. Goldblatt, D., & Perraton, J. (1999). *Global transformations: Politics, economics and culture*. Stanford: Stanford University Press.

Hirsch, F. (1976). *Social limits to growth*. Cambridge, Harvard University Press.

IDP Education Australia, with Australian Education International (2001). *Comparative costs of higher education courses for international students*. Sydney, IDP.

Institute for International Education, IIE (2006) Data on international education in the United States. Accessed 12 December 2006 at: http://www.iie.org/

Institute for Scientific Information, Thomson-ISI (2006). Data on highly cited researchers, ISIHighlyCited.com. Accessed 10 April 2006 at: http://isihighlycited.com/

James, R., Baldwin, G. & McInnis, C. (1999). *Which university? The factors influencing the choices of prospective undergraduates*. Evaluations and Investigations Program, Higher Education Division, Department of Employment, Education and Training, Australian government. Accessed 5 October 2004 at: http://www.dest.gov.au/archive/highered/eippubs/99-3/whichuni.pdf

Kelo, M., Teichler, U., & Wachter (Eds.) (2006). Eurodata: Student mobility in European higher education. Bonn: Lemmens Verlags- & Mediengesellschaft.

Kenway, J., Bullen, E., Fahey, J., with Robb, S. (2006). *Haunting the knowledge economy*. London: Routledge.

Kaul, I., Grunberg, I., & M. Stern (Eds.) (1999). *Global public Goods: International cooperation in the 21st century*. New York: Oxford University Press.

Luijten-Lub, A., van der Wende, M., & Huisman, J. (2005). On cooperation and competition: A comparative analysis of national policies for internationalization of higher education in seven Western European countries. *Journal of Studies in Higher Education*, 9(2), 147-163.

Mahroum, S. (2001). Foreign scientific researchers in selected OECD EU countries: Data sources and analysis. In OECD, *Innovative people: Mobility of skilled personnel in national innovation systems*, 219-228. Paris: OECD.

Mandel, E. (1975). *Late Capitalism*. J. de Bres (transl.). London: Verso.

Marginson, S. (1993). *Education and public policy in Australia*. Cambridge: Cambridge University Press.

Marginson, S. (1997). *Markets in Education*. Sydney: Allen & Unwin.

Marginson, S. (2004). Don't leave me hanging on the Anglophone: The potential for online distance education in the Asia-Pacific region. *Higher Education* Quarterly, 58(2/3), 74-113.

Marginson, S. (2007a). The public/private division in higher education: a global revision, Higher Education, 53, pp. 307-333

Marginson, S. (2007b). *'One little piece of endless sky... ' Global flows and global field: Imagining worldwide relations of power in higher education*. Seminar on 'Geographies of knowledge/geometries of power', University of Cardiff. Gregynog, Wales, 5-7 February. Accessed 18 June 2007 at: http://www.cshe.unimelb.edu.au/people/staff_pages/Marginson/ESRC_seminar0602067.pdf

Marginson, S. (2007c). Global position and position-taking: the case of Australia. *Journal of Studies in International Education*, 11(1), 5-32

Marginson, S. (2008). The academic professions in the global era, in J. Enders & E. de Weert (Eds.), The academic profession and the modernization of higher education: Analytical and comparative perspectives. Dordrecht: Springer [in production]

Marginson, S. & Considine, M. (2000). *The enterprise university: Power, governance and reinvention in Australia*. Cambridge, Cambridge University Press.

Marginson, S. & Ordorika, I. (2007). 'El central volumen de la fuerza'. Global hegemony in higher education and research. Paper prepared for the Social Sciences Research Council, New York. Accessed on 30 June 2007 at: http://www.ssrc.org/

Marginson, S. & Rhoades, G. (2002). Beyond national states, markets and systems of higher education: A glonacal agency heuristic *Higher Education*, 43, 281-309.

Marginson, S. & Sawir, E. (2005). Interrogating global flows in higher education. *Globalization, Societies and Education*, 3(3), 281-310.

Marginson, S. & Sawir, E. (2006). University leaders' strategies in the global environment: A comparative study of Universitas Indonesia and the Australian National University. *Higher Education*, 52, 343-373.

Marginson, S. & van der Wende, M. (2007). *Globalization and Higher Education*. Paper prepared for Organisation for Economic Cooperation and Development. Paris: OECD. Accessed 31 May 2007 at: http://www.cshe.unimelb.edu.au/people/staff_pages/Marginson/Marginson&van_der_Wende2006OECD.pdf

Marx, K. (1981). *Capital, Volume 3*. Harmondsworth: Penguin.

Mazzarol, T., Soutar, G., Smart, D., & Choo, S. (2001). *Perceptions, information and choice*. Canberra, Australian Education International. Accessed 25 May 2002 at: http://www.dest.gov.au

McBurnie, G. & Ziguras, C. (2006). *Transnational education* .London: Routledge.

McLuhan, M. (1964). *Understanding media*. Abacus, London.

Moogan, Y., Baron S. & Harris, K. (1999). Decision-making behaviour of potential higher education students. *Higher Education Quarterly*, 53(3), 211-228.

Musselin, C. (2004). Towards a European academic labour market? Some lessons drawn from empirical studies on academic mobility. *Higher Education*, 48, 55-78.

Musselin, C. (2005). European academic labour markets in transition. *Higher Education*, 49, 135-154.

National Science Board, NSB (2006). *Science and engineering indicators 2004*. Accessed on 9 April 2006 at: http://www.nsf.gov/statistics/seind04/

Organisation for Economic Cooperation and Development, OECD (2002). *International mobility of the highly skilled*. Paris, OECD.

Organisation for Economic Cooperation and Development, OECD (2003). Unpublished data on international education. Paris: OECD.

Organisation for Economic Cooperation and Development, OECD (2004a). *Internationalisation and trade in higher education*. Paris, OECD.

Organisation for Economic Cooperation and Development, OECD (2004b). *Quality and recognition in higher education: The cross-border challenge*. Paris: OECD.

Organisation for Economic Cooperation and Development, OECD (2004c) *OECD science, technology and industry outlook*. Paris: OECD.

Organization for Economic Cooperation and Development, OECD (2005a). *E-learning in tertiary education: Where do we stand?* Paris: OECD.

Organisation for Economic Cooperation and Development, OECD (2005c). Website of OECD on internationalisation of tertiary education. Accessed 11 September 2005 at: http://www.oecd.org/document/47/0,2340,en_2649_34859749_31392495_1_1_1_1,00.html

Organisation for Economic Cooperation and Development, OECD (2006). *Education at a glance*. Paris, OECD.

Pusser, B. (2002). Higher education, the emerging market and the public good. In P. Graham & N. Stacey (Eds.) *The knowledge economy and postsecondary education*. Washington: National Academy Press.

Pusser, B. (2003). Beyond Baldridge: Extending the political model of higher education organization and governance. *Educational Policy*, 17(1), 121-140.

Rhoads, R. & Torres, C. (2006). The university, state, and market: The political economy of globalization in the Americas. Stanford: Stanford University Press.

Samuelson, P. (1954). The pure theory of public expenditure. *Review of Economics and Statistics*, 36(4), 387–389.

Shanghai Jiao Tong University Institute of Higher Education, SJTUIHE (2007). Academic Ranking of World Universities. Accessed 15 March 2007 at: http://ed.sjtu.edu.cn/ranking.htm

Sidhu, R. (2005). Universities and globalization: To market, to market. Lawrence Erlbaum Associates.

Singh, M. (2001). *Re-inserting the 'public good' into higher education transformation*. Conference on Globalisation and higher education - views from the South, University of Capetown, March. South African Council on Higher Education: Pretoria

Smeby, J.-C. & Trondal, J. (2005). Globalization or Europeanization? International contact among university staff. *Higher Education*, 49, 449-466.

Stiglitz, J. (1999). Knowledge as a global public good. In I. Kaul, I. Grunberg, & M. Stern (eds.) *Global public goods: International cooperation in the 21st century*, 308-325. New York: Oxford University Press.

Taskforce on Higher Education and the Developing World (2000). *Peril or promise*. Report. Washington: World Bank.

Teese, R. (2000). Academic success and social power. Melbourne: Melbourne University Press.

Teichler, U. (2004). The changing debate on internationalization of higher education. *Higher Education*, 48, 5-26.

Teixeira, P., Jongbloed, B., Dill, D., & Amaral, A. (Eds.) (2004). Markets in higher education: Rhetoric or reality? Dordrecht: Kluwer

Times Higher (2007) World university rankings, *The Times Higher Education Supplement*. Accessed on 15 March 2007 at: www.thes.co.uk

Tremblay, K. (2005). Academic mobility and immigration. *Journal of Studies in International Education*, 9(3), 196-228.

U.S. News and World Report (2006). *America's best colleges, 2007 edition*. Washington: USNWR.

Vincent-Lancrin, S. (2006) What is changing in academic research? Trends and futures scenarios. Draft paper. Paris: OECD-CERI.

Winston, G (2003). *Towards a theory of tuition: Prices, peer wages, and competition in higher education*. Discussion Paper Number 65, Williams Project on the Economics of Higher Education. Williamstown: Williams College.

Woolcock, M. (2001). The place of social capital in understanding social and economic outcomes. In Human Resources Development Canada and OECD, *The contribution of human and social capital to sustained economic growth and well-being*, International Symposium Report, September, 65-88. OECD: Paris.

WTO/GATS (2005). World Trade Organization website on negotiations on the General Agreement on Trade in Services (GATS) in relation to educational services. Accessed 11 September 2005 at: http://www.wto.org/english/tratop_e/serv_e/education_e/education_e.htm

World Bank (2007) World Bank data and statistics, Accessed 28 February 2007 at: http://www.worldbank.org/data

Simon Marginson
Centre for the Study of Higher Education
University of Melbourne
Melbourne, Australia

SIMON MARGINSON

3. GLOBAL UNIVERSITY RANKINGS

INTRODUCTION: IDEALIZED MODELS VS. GLOBAL DIVERSITY

There is no one single 'Idea of a University' (Newman 1899/1996). There are many different missions, structures and organizational cultures, associated with distinctive traditions and models of higher education. All of the multiple 'Ideas' are nested in national contexts, historical identities and conditions of possibility. In the 'Westminster' countries (UK, Australia, New Zealand) national systems combine university autonomy with explicit central steering. The Nordic/ Scandinavian university combines emphases on high participation social equity, research culture and university autonomy with strong state investment (Valimaa, 2004; 2005). The German-style university opts for elite participation, research culture and state administration. The Latin American public university has high participation, scholarly culture and a central social and political role in the building of the nation-state. The emerging science universities of East and Southeast Asia, including China, Taiwan, Korea and Singapore are fostered by state investment and in Singapore are explicitly designed to secure global competitiveness. India has technology and business-focused institutions that at best combine high quality and commercialism. Beyond the research university are strong vocational sectors in Finland, Germany (the Fachhochschulen), France, and other vocational and community-based programs, not to mention the different online institutions. There are also many specialized institution in teaching and research.

But some traditions are more equal than others. In the templates of the New Public Management (NPM), which has spread to higher education systems through most of the world,[27] the pace of its distribution quickened by policy globalization including its adoption by the World Bank in lending programs in the developing world, the favoured models are taken from the United States, or rather, a reified, idealized version of American higher education as a national market.

The NPM norms higher education in terms of government-steered competition between institutions, executive-steered competition between academic units, systems with a mixture of public and private institutions, institutions with a mixture of public and private funding, higher tuition, customer focus; university marketing and entrepreneurship, research links with industry; performance measures and output-based funding; and relations with funding agencies based on contracts, accountability and audit. NPM reforms are driven by desires for fiscal

[27] Numerous studies, supportive and critical, attest to the impact of the New Public Management in system organization and institutional management: for example Clark, 1998; Marginson & Considine, 2000; Nowotny et al., 2001; Musselin, 2005; Henkel 2005, 2007; Rhoads & Torres, 2006.

Simon Marginson (ed.), Prospects of Higher Education, 79–100.

efficiency, global competitiveness and more effective control systems. The NPM sees national systems as economic markets and imagines institutions as firms driven fundamentally by economic revenues and market share, not teaching, research and service as ends in themselves. US higher education, which is underpinned by nationally financed student loans that enable a relatively high level of mobility, is traditionally understood as a national market.

However, there is more than one possible 'American model' in a national system that is a by-word for structured diversity, as summarised in the Carnegie classification (McCormick & Zhao, 2005). There are the Ivy League private research university and the liberal arts college, the state universities and four year institutions, the community college and the newer model of for-profits trading on the equity market. Most US enrolments are in the public sector, which includes the University of California system and the land grant and other public flagship universities in different states (Turner & Pusser, this volume). But the US system is also a by-word for hierarchy and some classifications have higher status. The NPM has taken two institutional norms or models from the US experience for world-wide propagation. These are not so much blueprints as broad social imaginaries diffused across nations, institutions and social agents; often in vague forms and imprecise notions; and combining structures, technologies, behaviours and values:

> *The idealized model of research university*, the 'entrepreneurial model' (Clark, 1998): centrally focused on knowledge production; emphasizing research and graduate studies, excellence and prestige; tied to business and the knowledge economy; competitive for students and funds; internationally oriented, achieving greater autonomy via financial diversity, including tuition and philanthropy. The US Ivy League private universities are closest to this model in the real world and while they are not as consumer responsive or efficient as the model imagines they impart to it a tremendous prestige;

> *The idealized model of the for-profit vocational university*: focused on vocational training for business, computing, perhaps the mass professions such as health; using teaching from the vocational field and curriculum packages, and accountable for its immediate vocational relevance, marketing-driven in its business planning and organizational culture; relentlessly expansionary in student numbers, sites and market share; spare and efficient with no 'frills' such as research, libraries or academic freedoms; 'customer'-focused using performance management of staff and quality assurance. This form has a mixed record around the world but in the US it has become embodied in corporations that raise significant equity funds (Ortmann, 2002), including the Apollo Group, parent company of the University of Phoenix. Phoenix is the largest and fastest growing private university in the USA and has spread to a dozen other countries, although in 2007 it ran into difficulties.

It is ironic these models depart from actual American practice in significant ways. First, US higher education is more politicized and less driven by pure market forces than the NPM imagines. Second, private institutions are dependant on state support via student loans, and in the case of the Ivy League the public funding of research. Public subsidies are essential to the global strength of all US institutions. Third, neither model fits the majority of American institutions, which as noted are in the public sector. This has not stopped these two models from shaping not only the policy imagination in many countries, but also shaping the emerging global market, grounded in cross-border positional goods and the stratification of research reputation as Chapter 2 described. The idealized research university is the 'Super-league' Anglo-American research university in the elite sub-field of the global market. The idealized for-profit vocational university is the perfected model of the mass education and commercial providers of cross-border degrees at the other end of the global market. To restate Anderson's (1983) point about the nation as an 'imagined community', an emerging market is an 'imagined competition'.

University rankings. Thus hegemonic idealized models become turned into living breathing social systems. In the process of normalizing cross-border education as a global market, the most potent mechanism has been university rankings, which began in 2003 (SJTUIHE, 2007; Times Higher, 2006). Global university rankings are tools of the imagination, discursive technologies and data sets whose time has come. But university rankings also frame the imagination, and make the time.

Notwithstanding their imagined (not to mention limiting, reductive and reifying) qualities, as the above account indicates normative global university rankings have not come from nowhere. They are grounded in a particular kind of prestigious institution which has helped to give them purchase in the sector; and nested in history. Hierarchical global comparison has been made possible by a particular kind of global condition in which the groundwork for imagining higher education as a universal competition had been laid not only by global economics, culture and communications and by the NPM; but more specifically in higher education by the normative methods of comparative education and development education (Marginson & Mollis, 2001); and also by a history of national rankings in different nations, especially the *US News and World Report* (2006). At the same time, the practices of global university rankings, which have gripped the attention of higher education institutions, governments, media and public throughout much of the world, also tend to more deeply entrench that global condition and legitimate that process of universal cross-border comparison. Rankings have become an end in themselves, without much regard as to exactly what they measure, whether they are solidly grounded or whether their use has constructive effects. The desire for rank ordering overcomes all. League tables rule. (Often institutions are rank ordered even where differences in the data are not statistically significant). This frees up the endogenous logic of rankings to reshape the sector. It recalls Foucault in *The archaeology of knowledge* (1972), where he describes discourses as 'practices the systematically form the objects of which they speak' (Foucualt, 1972, p. 49).

There is also a more immediate politics of rankings. Different rankings systems have particular ways of recording, classifying and counting, with specific shaping effects. These effects have direct implications for the rise and fall of actual institutions, not to mention whole national systems, in cross-border relations and in the distribution of prestige in the home country. Outside the USA, where only national rankings really matter, the standing of a university is boosted by a good global position. In other words rankings further the specific interests of certain kinds of institution, and particular institutions and nations (though these effects can differ between the different rankings systems). Rankings not only imagine and describe higher education in terms of Bourdieu's notion of a universal positional competition in the sector, they engender precisely the kind of competitive struggle that he discusses: a struggle over social position itself, and over the terms in which positionality is understood, described and fixed. Many institutions attempt to influence the system of classification and the contents of the data input used in the rankings systems, though in this some are more effective than others. It recalls Bourdieu's comment in *Distinction: A social critique of the judgment of taste* (1984) that: 'Principles of division, inextricably logical and sociological, function within and for the purposes of the struggle between social groups; in producing concepts, they produce groups, the very groups which produce the principles and the groups against which they are produced' (Bourdieu, 1984, p. 479).[28]

This chapter. The chapter outlines the principal global rankings systems in use and some issues and problems that have arisen, and discusses their effects in the policies of governments and institutions. It then examines critiques of the existing rankings systems, and alternative systems of comparison and classification that have emerged, mostly in Europe, before an overall summation.

THE FIRST GLOBAL RANKING SYSTEMS

National Rankings

Most countries with large higher education systems have rankings of one kind or another. Countries with rankings devised by newspapers and magazines include China and Hong Kong China, Japan, India, the Ukraine, Romania, Poland, Portugal, Italy, Spain, Germany, Sweden, Switzerland, France, UK, USA and Canada. In Thailand, Malaysia, Pakistan, India, Kazakhstan, Korea, Tunisia, Nigeria, the Netherlands, the UK, Brazil and Argentina rankings have been instigated by ministries of education, grants councils or accreditation agencies. In China, Japan, Australia, Kazakhstan, Slovakia, Romania, Russia, the Ukraine, Germany, Spain, Switzerland, the UK and Canada rankings have been initiated by universities, professional associations or other organizations (Salmi & Saroyan, 2006). Usher and Savino (2006) review rankings complied in Australia (Williams & Van Dyke, 2006); Canada; China, with several rankings systems (Liu &d Liu, 2005), and Hong Kong; Germany; Italy; Poland; Spain; the UK, where four different newspapers have developed rankings systems; and the USA.

[28] In the same work Bourdieu remarks pessimistically (or is it optimistically?) that 'the whole process constitutes a perfect circle from which the only escape is to objectify it sociologically' (Bourdieu, 1984, p. 235).

There are also many specialist rankings systems that focus on characteristics ranging from research output, to student services, to MBA programs, to Yahoo Magazine's ratings of 'connectivity', the university contribution to social diversity and other features. Specialist rankings are not discussed further in detail here but it can be noted that ratings of disciplines and other aspects are also taken up in the larger schema of the Shanghai Jiao Tong University and the German CHE.

US News and World Report rankings: For the most part national rankings consist of a single table but in the USA and Canada higher education institutions are divided into groups according to mission and other characteristics, creating a set of mini league-tables within the rankings system. In the United States the annual *US News and World Report* (USNWR) survey focuses principally on aspects of institutions seen to contribute to the quality of teaching and the student experience, thereby servicing the national market of choice-making students whose potential mobility is underpinned by national loans and in the case of selected students, the broad range of tuition scholarships and other subsidies provided by institutions. There is less emphasis on research and scholarship. The current categories of institutions are drawn from the classification in 2000 by the Carnegie Foundation for the Advancement of Teaching (USNWR, 2006, p. 81ff). The most important grouping is the 'National Universities', 248 universities (162 public and 86 private) each providing a wide range of fields of study and offering degrees to doctoral level. Most of these institutions are research intensive. Other categories of institution used by USNWR are liberal arts colleges, Master's level universities, Comprehensive Bachelor level universities, business programs, undergraduate engineering programs, and speciality schools, mostly in fine and performing arts.

In the categories of national universities and liberal arts colleges, the largest part of the index (25 per cent) is comprised by a survey of 'undergraduate academic excellence' sent to university presidents, provosts and deans of admissions. Two items each constitute 20 per cent: student retention and graduation rate; and 'faculty resources' which rewards small classes, high academic salaries, high academic qualifications and a high proportion of staff full-time. Student selectivity at entry, a proxy for positional market demand, is 15 per cent. The lesser items are spending per student (10 per cent), the proportion of alumni who donate back to the institution (5 per cent), and the graduation rate after controlling for spending and student aptitude (5 per cent) (USNWR, 2006, pp. 77-79). These details have become very significant to institutions. Over the years institutions have learned to modify and target their behaviour so as to maximise their USNWR position. Arguably, this has exacerbated the reputational race (Frank, 2001). It has also had more specific effects that are perverse from the public interest viewpoint, for example the manipulation of student entry to maximise student scores and refusal rates, and the growth of merit-based student aid at the expense of needs-based aid to improve student selectivity, retention and graduation indicators. In *Shakespeare, Einstein and the bottom line* (2004) Kirp provides case studies on these practices.

Table 3.1. Shanghai Jiao Tong university rankings: weightings

	weighting
Alumni of institution: Nobel Prizes and field medals	10%
Staff of institution: Nobel Prizes and field medals	20%
High citation (HiCi) researchers	20%
Articles in citation indexes in science, social science and humanities	20%
Articles in Science and Nature	20%
Research performance (as above) per head of staff	10%
total	100%

Source: SJTUIHE, 2007

The Jiao Tong University Global Ranking

The first world-wide ranking by the Shanghai Jiao Tong University Institute of Higher Education (SJTUIHE) was in 2003. The SJTUIHE rankings process was supported by the Chinese government, which wanted to be able to assess the research performance of Chinese universities according to accepted international standards, as one basis of strategies designed to lift the global competitive position if science and technology R&D in China. Institutional rankings have been issued annually since, and in February 2007 and rankings in five broad disciplinary fields were released. The sole focus of Jiao Tong attention is research. The Jiao Tong group argues that the only data sufficiently reliable for ranking purposes are broadly available and internationally comparable data of measurable research performance (Liu & Cheng, 2005, p. 133). It is considered impossible to compare teaching and learning 'owing to the huge differences between universities and the large variety of countries, and because of the technical difficulties inherent in obtaining internationally comparable data'. Further, the Jiao Tong group states that it does not want to employ subjective measures of opinion or data sourced from universities themselves, as are used in some national rankings (SJTUIHE, 2007). The Jiao Ting group consults widely and continually in relation to data compilation and interpretation, and holds an annual conference of rankings experts in Shanghai. Over this time the measures have been developed, refined and tightened so that within its framework the rankings system has been successively strengthened.

As Table 3.1 shows the Jiao Tong rankings are based on a composite index whereby the weighting given to different criteria is determined by the rankings agency index. The index is primarily driven by academic publication and citation, principally in science-based disciplines with some attention to social sciences and humanities: 20 per cent citation in leading journals as listed by the publishing company Thomson; 20 per cent articles in *Science* and *Nature*; and 20 per cent the number of Thomson/ISI 'HiCi' researchers in the top 250-300 in their fields on the basis of citation (ISI-Thomson, 2006). Another 30 per cent is determined by the winners of Nobel Prizes in the sciences and economics and Fields Medals in mathematics, based on the location of training (10 per cent) and current employment (20 per cent). The remaining 10 per cent is determined by dividing the total derived from the above data by the number of academic staff eligible to conduct research. The separate discipline rankings fall into five categories: natural sciences and mathematics, engineering/technologies and computer sciences, life and agricultural sciences, clinical medicine and pharmacy, and social sciences. While the same broad schema is followed in the determination and weighting of criteria, there is some variation by discipline due to publishing conventions. (Full details of each method of discipline ranking are provided at SJTUIHE, 2007).

The Jiao Tong index is methodologically sound and for the most part provides a valid basis for synchronic global comparisons. It measures only real outputs, rather than subjective assessments of reputation which may or may not be solidly grounded. Its methods are transparent. The exception is the Nobel Prize and Fields Medal criterion. Nobel Prizes are submission based; science and scholarship are not the only determining factors. Politicking can enter the decisions. Regardless of questions of methodological soundness however, the SJTUIHE rankings value universities closest to the model of institution that the designers wished to emulate The Jiao Tong calculations favour universities large and comprehensive enough to amass strong research performance over a broad range of fields while carrying few research inactive staff. They favour universities very strong in the sciences, universities from English language nations because English is the language of research (non English language work is published less and cited less) and universities from the USA because Americans tend to cite Americans. The number of Thomson/ISI 'HighCi' researchers directly and indirectly drives performance in much of the index. As noted in chapter 2 the HiCi researchers are overwhelmingly located in the USA, with more than eight times as many as the next nation, the UK (ISI-Thomson, 2006); Jiao Tong research performance is dominated by the English speaking nations, with 71 per cent of the world's top 100 research universities in 2006, particularly by the United States with 17 of the top 20 and 54 of the top 100.

There are clear winners and losers but nevertheless, the Jiao Tong rankings have arranged world-wide higher education in a hierarchy based on a technically defensible measure of performance. As such they directly define the elite sub-field of the global market. The only doubt is where the border falls. Strictly, this is impossible to resolve in technical terms, but by naming a top 100 with place-by-place rankings and an overall top 500 inclusions the SJTUIHE provides a two tier descriptor of the global research hierarchy. (As noted in chapter 2, this does not alter the fact that USA and perhaps UK universities in the top 100 enjoy a higher status than their counterparts from elsewhere). The Jiao Tong rankings do not rank universities according to student demand or volume in the cross-border degree market. The SJTUIHE group has been at pains to emphasise that this is not a holistic ranking of universities in all respects, including teaching quality, it refers only to research performance. But the impact is broader. By arranging universities according to research status, the primary determinant of the value of the positional goods produced in both the national markets and cross-border programs, the Jiao Tong rankings construct a reputational hierarchy. Harvard becomes understood not as number one research site but as number one university. Market research and anecdotal evidence from educational agents indicate that the Jiao Tong rankings are feeding directly into student choice-making in relation not only to doctoral programs, which are nested in research, but first degree and Masters programs as well. It is as significant to be excluded from the Jiao Tong top 500 as to be ranked within it. De facto the SJTUIHE has constructed a market across the entire sector.

Effects on behaviour. Except in the USA there have been profound effects on behaviours and policies. The Jiao Tong initiative has triggered many conferences on ranking, in which the SJTUIHE group itself was active, and a flurry of marketing by the successful institutions. More generally, it has catalysed in each university a desire to improve its league table position by affecting data compilation, or by improving research performance, or both. At government level, the Jiao Tong rankings have triggered the desire to see more universities in the upper echelons of the table, as a signifier of national success and potential in the policy context of a global knowledge economy. In turn this has led to policies designed to concentrate a higher proportion of research activity and resources in a selected group of leading universities, and to attract and hold more top flight researchers. Singapore has set out to attract high quality expatriates by paying American level salaries. China and Germany have selected a group of universities for special research status (van der Wende, 2007a). Others, such as Australia and the UK[29] are talking about it. The strategic objective is to replicate the global attractiveness and impact of the American doctoral sector, though in the absence of US economic, technological, political and cultural power this is difficult to do. In Europe, the accelerated research race is leading to multilateral action. Leiden University in the Netherlands has led the formation of a League of European Research Universities; the European Union plans a European Institute of Technology (EIT) (van der Wende, 2007a; 2007b). At the time of writing it was unclear whether the EIT would be based on existing institutions or be a new institution with virtual and/or corporeal character.

The Jiao Tong rankings trigger a positive relationship between ranking measures and total research activity, providing institutions have the policy autonomy and resources to introduce the internal reforms necessary to lift relative position (Salmi & Saroyan, 2006, p. 23). For individual institutions the benefits are more doubtful. Non research institutions are pushed outside the positional hierarchy. For research universities there are feedbacks between research ranking, investment and activity; but with all universities striving for the same goal within a zero sum game, an average effort leaves the individual institution no better off than before. As always happens when positional competition is intensified, if everyone stands on tip-toe no one can see better. 'The race gets longer for the same prize' (Hirsch, 1976, p. 67).

29 The UK's increased focus on concentration policy post-Jiao Tong is curious given that it already has a successful concentration program. History, the Research Assessment Exercise (RAE), and the distribution of targeted research-specific funding associated with the RAE have together produced a stratified allocation of capacity. The Russell Group of universities is already functioning as an elite research sub-sector; and Cambridge and Oxford are included in every global top ten, with Imperial and University College London just below. The UK has 11 universities in the Jiao Tong top 100 (the USA has five times as many but spends 17 times as much as does the UK on higher education) suggesting that the UK is doing relatively well in terms of measured research performance.

Table 3.2. Times Higher university rankings: weightings

	weighting
'Peer review' survey of academics	40%
Survey of 'global employers'	10%
Proportion of academic faculty who are foreign	5%
Proportion of students who are foreign	5%
Staff-student ration (proxy for 'teaching quality')	20%
Research citations per head of academic faculty	20%
total	100%

Source: THES, 2006

The Times Higher Global Ranking

The first *Times Higher Education Supplement* 'World University Rankings' were published in 2004. The aim is a summative, holistic ranking of university 'quality'. Again the index is composite, and also heterogeneous, covering much more than research. Half the Times Higher index is grounded in existing reputation, with 40 per cent comprised by a survey of academics ('peer review') and 10 per cent by a rsurvey of 'global employers'. There are two internationalisation indicators: the proportion of students who are international (5 per cent) and the proportion of staff (5 per cent). Another 20 per cent is fixed by the student-staff ratio, a quantity measure used as a proxy for teaching 'quality'. The other 20 per cent is research citation performance per academic staff. The Times issues an annual list of the top 200 universities in order of merit, plus rankings by institution in natural sciences, engineering and IT, biomedicine, social science and arts/humanities (THES, 2006).

Methodologically, the 'Times tables' are open to criticism, in relation to most components of the index. In relation to the two reputational surveys, it is not publicly specified who is surveyed or what questions are asked. In 2006 the survey gathers a response of just 1 per cent from 200,000 e-mails sent worldwide. Not all responses were valid and could be used, and a relatively high proportion were from nations where the *Times* is well-known (Sowter, 2007), so that responses from the UK, Australia and former British colonial locations in Southeast Asia tended to be over-represented. Despite this composition effect the pool of responses were not re-weighted to create global evenness (probably impossible given that with an overall response is 1 per cent some cells had too few responses for re-weighting). In addition the published data for 2006 amalgamated the responses to the 2004, 2005 and 2006 surveys, despite variations in method across the three years. The 2004 and 2005 surveys also experienced low response and coverage bias. In addition the student internationalization indicator rewards entrepreneurial volume building but this is not necessarily grounded in the quality or quantity of student demand or the quality of programs or services. Nor can teaching quality be adequately assessed using student-staff ratios. Research plays a lesser role in the Times index than in most other comparisons of universities The Times Higher rankings reward a university's marketing division better than its researchers.

In composing the global market the Times Higher rankings are more ambiguous than those of the SJTUIHE. Leaving aside composition bias, on one hand the 50 per cent of the index measuring comparative reputation reproduces the established global hierarchy. Leading research intensive universities are on top. This effect is reinforced by the 20 per cent of the index derived from research performance. At the same time, reputation is itself a composite category, taking in more than just comparative research performance. Elements such as national prestige, the standing of universities within the national system, and marketing, also come into play. Compared to the Jiao Tong ranking, the Times Higher ranking tends to elevate well known institutions in national systems especially those in national capitals. It is likely that the Times Higher ranking has also been influenced by the relatively heavy self-marketing and promotion by some Australian and UK universities in the cross-border student market. Likewise the internationalisation of students indicator rewards institutions active in the cross-border student market. Because it is a quantity indicator, it unambiguously rewards a mass orientation within that market. Hence the Times Higher combines, in eclectic fashion, both of Bourdieu's principles of hierarchization. The result is a less coherent hierarchy.

Many universities that have done well in the Times Higher rankings have not hesitated to exploit the marketing benefits, particularly universities active in cross-border marketing. Institutions have also been vigorous in lobbying QS Marketing, the company that carries out the data collection process, which unlike the Jiao Tong group uses data collected from the institutions themselves as well as sources such as government-generated statistical collections. However, the other effects on institutional behaviour and national policy are less clear. The ambiguity of the Times Higher rankings (including the weaknesses in methodology and data collection) mean that there is no clear line of cause and effects between improved performance and improved ranking position. The Times Higher rankings are less definitive, less defensible, more criticised and less influential than the Jiao Tong. In particular, the data series has been negatively affected by two factors: the volatility of the rankings, particularly below the leading group of universities, and what appears to be the over-performance of UK and Australian universities which in part is an effect of the composition bias in the survey of academic peers.

Volatility of the Times Higher rankings. In 2004 the oldest public university in Malaysia, the University of Malaya was ranked by the Times Higher at 89. The newspapers in the Malaysian capital, Kuala Lumpur, celebrated. The Vice-Chancellor ordered banners declaring 'UM a world's top 100 university' placed around the city, with one on the edge of the campus facing the main freeway to the airport where every foreign visitor to Malaysia would see it. But the next year in 2005 the identity of Chinese and Indian students at the University was correctly shifted from international to national, and there were changes in other parts of the Times composite indicator. The University of Malaya dropped from 89 to 169. In Malaysia it seemed as if the University's reputation abroad and at home was in free fall. The VC was pilloried in the local media, and when his position came up for renewal by the government in March 2006 he was replaced.

Table 3.3. Some volatile movements in the Times Higher rankings, 2004-2006

University	*Times Higher ranking*		
	2004	*2005*	*2006*
Fudan, China	195	72	=116
Tsing Hua, China	61	= 62	28
Keio, Japan	-	215	120
Kyushu, Japan	-	222	=128
Osaka, Japan	69	=105	= 70
Tokyo Institute of Technology, Japan	51	99	118
Seoul National, Korea	118	= 93	63
Malaya, Malaysia	89	=169	192
Chulalongkorn, Thailand	-	=121	=161
Royal Melbourne Institute of Technology, Australia	55	82	146
Queensland Technology, Australia	-	118	=192
Massey, New Zealand	108	=188	-
Otago, New Zealand	114	=186	= 79
Munich, Germany	99	55	98
Ecole Pol. Federale de Lusanne, Switzerland	32	34	= 64
Geneva, Switzerlands	-	= 88	39
Basel, Switzerland	-	=127	75
Ecole Polytechnique, France	27	10	37
Sciences Po, France	-	69	52
Birmingham, UK	126	=143	= 90
School of Oriental & African Studies, UK	44	=103	= 70
Dartmouth, USA	138	117	= 61
Emory, USA	173	141	56
Pittsburgh, USA	-	193	88
Purdue, USA	59	61	127
Vanderbilt, USA	156	=114	53

Source: THES, 2006 and predecessors

It was not just the University's executive leader who was subordinated by the reclassification; the reputation of the University itself (long established and one of two strongest research universities in Malaysia) and the national research system were equally affected. Yet the University of Malaya had dropped 80 places without any decline in its real performance. The same kind of problems has occurred in many other universities, as Table 3.3 shows. Sharp rises and falls especially affect the second half of the top 200 where small differences in metrics generate large rankings effects. This pattern of sharp changes in the position of a large minority of universities from year to year contrasts with the more stable Jiao Tong rankings.

Composition bias. Both the UK and Australia do exceptionally well in the Times Higher rankings. British universities have improved their comparative performance each year. The UK has 15 per cent of the GDP of the USA but almost half as many universities in the Times Higher top 100: 15 in the UK and 33 in the USA in 2006. In comparison, the USA has 54 research universities in the Jiao Tong top 100.

Table 3.4. Australian universities in the Times Higher ranking of the top 200, 2004-2006

University	*Times Higher ranking*		
	2004	*2005*	*2006*
Australian National University	18	23	16
University of Melbourne	22	19	22
University of Sydney	40	=38	=35
Monash University	33	33	38
University of New South Wales	36	40	41
University of Queensland	49	47	45
Macquarie University	68	67	=82
University of Adelaide	56	=80	=105
University of Western Australia	96	=80	=111
Royal Melbourne Institute of Technology	55	82	146
Curtin University of Technology	76	=101	=156
Queensland University of Technology	--	118	=192
University of Wollongong	--	--	196
La Trobe University	142	98	--
University of Newcastle	--	=127	--
University of South Australia	--	=154	--
University of Tasmania	161	=166	--

Source: THES, 2006 and predecessors -- indicates ranking position outside the top 200

Remarkably, the Times has succeeded in reducing the American hegemony from 54 to just 33 of the top 100. The British now have two of the Times Higher top three universities and in the 2006 results Cambridge UK has almost closed the gap on Harvard. Yet the Harvard faculty is cited at three and a half times the rate of its British counterparts, and arguably has a much stronger global reputation. This result appears to be an artefact of the composition bias in the peer survey.

Likewise, the performance of the Australian universities also appears to have been inflated. Despite a relatively poor citation rate and moderate staffing ratios they do exceptionally well in the peer survey of academic staff which comprises 40 per cent of the index, and the internationalisation indicators especially that for students. In 2006 the Australian National University and the University of Melbourne were both ranked by the 'academic peers' on the same level as Yale and ahead of Princeton, Caltech, Chicago, Columbia and Penn. Australia has 13 of the Times top 200 (Table 3.4) making it the third strongest system in the world, ahead of Japan, Canada, Germany and all of Western Europe. This makes sense in relation to international marketing but not all round performance or reputation.

CRITIQUES AND ALTERNATIVES

The potency of global rankings, unease with the effects and the fact of as many losers as winners, has generated an industry of critique. This ranges from criticisms conducted within the terms of the exercise, to those that look beyond it to the larger open global setting (Marginson & van der Wende, 2007; chapter 2 this volume).

Normalization and diversity. The first and principal concern is the implications of normalizing models on the actual existing diversity in higher education. Here the criticisms are mounted more against bias effects than long term reshaping effects and to that extent are conducted within the terms of the exercise. Bias includes the favouring of English language countries, and comprehensive universities in both rankings system and the favouring of science-based research intensive universities in the Jiao Tong rankings. One example is that of the hegemonic national public university in Latin America. Each of Universidad Nacional Autonoma de Mexico (UNAM) and Universidad de Buenos Aires in Argentina (UBA) provide access to more than 200,000 students on many sites and perform many functions in national and regional development, and social and cultural life, including national research leadership. This range prevents UNAM and UBA from concentrating resources so as to maximise research intensity and reputation like a Princeton or a Caltech, and scholarship in Spanish is unrecognised in the rankings process. They appear in the Jiao Tong rankings in the 151-200 bracket. UNAM is 74 in the Times Higher table. Yet these universities are manifestly great, and the model has distinctive strengths and a long international role. Further, as noted, beyond the research university are other kinds of institutional mission, from liberal arts colleges in the USA to the many institutions focused on vocational and professional training without a role in basic research. The reputational element in the Times Higher ranking allows it to include the Indian IITs and elevate the Dutch technical research universities above traditional institutions but vocational sectors such as the German Fachhochschulen remain excluded. In Europe concerns about diversity have triggered the evolution of institutional typologies for separated comparison, as discussed below.

On the question of language bias Altbach (2006) remarks that:

> The fact is that essentially all of the measures used to assess quality and construct rankings enhance the stature of the large universities in the major English-speaking centres of science and scholarship and especially the United States and the United Kingdom (Altbach, 2006, p. 3).

English is one of two languages spoken by approximately one billion people. The other is Putonghua ('Mandarin' Chinese). Two pairs of related mutually intelligible languages are spoken by over half a billion people: Hindi/Urdu (900 million) and Spanish/Portuguese (450/200 million). There are Russian (320 million), Arabic (250 million), Bengali (250 million) and four more languages with 100 million plus speakers: Malay-Indonesia, Japanese, French, German (Linguasphere Observatory, 2006). These languages and others are unlikely to disappear. But knowledge in other languages *has* disappeared from global rankings. Most academics in English language nations are monolingual but this is not the case elsewhere. English is the only global language of research publication but not the only language of thought. The exclusion of work in languages other than English is most problematic in the humanistic social sciences and humanities/arts where national languages dominate. These problems have generated a range of responses in the non English-speaking world, from lobbying for the inclusion of non-English publications in rankings systems, to the creation of separate language-based lists, to rejection of rankings.

More generally, because rankings are a normalising device in which any and every institution is imagined as sufficiently similar to be fully comparable with each other, they encourage context-free judgements about merit. Once institutions are arranged in a single league table the particular historical, economic and cultural factors that shape relative performance are obscured. Rankings foster the illusion of a level playing field in which each institution can rise up the table on merit and its place is determined by itself. But all institutions have specific conditions of possibility, and global higher education is not a level playing field. Institutions are tied to their history, to local context, to national resourcing (especially for expensive basic research programs), to their language of use and to the capacities of their personnel. They have self-determination, within variable boundaries. All can improve, but none have absolute freedom to move up to any possible level.

Ironically, at the same time rankings discourage initiatives in mission and organization, curriculum, pedagogy and research, except for initiatives within a short distance of the normalising model. Rather, rankings encourage university executives to concentrate energy on maximising performance and reputation within the established criteria comprising the rankings, particularly research performance and student selectivity. They raise the opportunity cost of investment in innovations in curriculum, pedagogy and research. Such innovations typically take years to be realised in improved performance and longer to feed into reputation. In this manner rankings reify not only the historically-inherited diversity, but future possible diversity as well. The normalising effects of rankings are at their worst when there is only one or a very small number of rankings systems. The more diverse and multiple the comparisons the less potent the normalising effects of one.

Problems of coverage, aggregation and summation. Rankings systems based on institutions appear to 'evaluate universities as a whole' (van Dyke 2005, p. 106), as the Times Higher claims. But no rankings system can cover all purposes of higher education from the point of view of all stakeholders. The outputs are multiple and complex, and not all are open to numerical measurement. Any ranking system will encompass the needs of some stakeholders better than others. Usher & Savino (2006) examine 19 league tables and university rankings systems, noting that these are driven by different purposes and varied notions of what constitutes quality. When rankings systems purport to cover the generality of purpose and interest, the partiality and exclusion are hidden, and so compounded.

This goes also to the issue of aggregation and weightings: Usher & Savino (2006) remark on the arbitrary character of the weightings used to construct composite indexes covering different aspects of quality or performance, the means by which ratings agencies construct a singular descriptor of institutions ranked against each other. Generally, composite indicators and weightings are untheorised, and 'the fact that there may be other legitimate indicators or combinations of indicators is usually passed over in silence. To the reader, the author's judgment is in effect final' (Usher & Savino 2006, p. 3). As Salmi and Saroyan (2006, p. 9) note: 'the weightings vary across league tables and typically reflect the view of the publisher of the table rather than being grounded'.

The areas excluded by the Jiao Tong and Times Higher rankings include teaching quality. The observable quality of teaching is difficult or impossible to measure with rigour for comparative national purposes, let global comparison; and no ranking or quality assessment system has generated comparative data based on the 'value added' during the educational process (Dill & Soo, 2005, pp. 503 & 505). Few comparisons focus on teaching and learning. Various proxies for teaching 'quality' have been used such as student-staff ratios, student selectivity (really a proxy for reputation not teaching quality), and research performance. Some might assume that research performance is positively correlated to teaching quality but 'empirical research ... suggests that the correlation between research productivity and undergraduate instruction is very small and teaching and research appear to be more or less independent activities' (Dill & Soo, 2005, p. 507).

Problems specific to reputational rankings. While reputational rankings provide a sense of where market perceptions are at, they are problematic when used as the main tool or mixed with objective data. They are circular. Subjective reputation, reproduced in the rankings, becomes recycled as subjective reputation. Well known university brands generate 'halo' effects.[30] Newcomers are blocked regardless of merit. In any case many rankers are unfamiliar with institutions not their own.

There is no way to verify the soundness of the subjective judgements of reputation by ensuring they are grounded in actual comparative knowledge.

European Responses to the Problems of Global Rankings

Practical responses to the problems inherent in the holistic rankings developed by Shanghai Jiao Tong University and the Times Higher have centred on Europe. These responses have taken four forms. First, development of a comprehensive comparison of teaching and research based on survey data, by the Centre of Higher Education Development (CHE) in Germany. Second, construction of a range of publication and citation metrics by a Leiden University group. Third, negotiation of a typology of European institutions enabling a more diverse set of comparisons than permitted by single scales. Fourth, the development of an agreed set of principles to guide inter-institutional comparisons, the Berlin Principles.

The Centre for Higher Education Development (CHE) comparisons. The Centre for Higher Education Development (CHE), located in Gutersloh in North Rhine-Westphalia in Germany, has developed an approach to comparisons between the programs of different universities, in conjunction with the German Academic Exchange Service which assists international students and the publisher Die Zeit.

The CHE dispenses with holistic rank ordering of institutions in league tables, avoids problems of arbitrary weighting and composite indicators; admits multiplicity of purpose into the process of comparison, and has shifted the

[30] The classical case is the American survey of students that ranked Princeton law school in the top ten law schools in the country. But Princeton had no Law school (Frank & Cook, 1995, p. 149).

normative power of the comparison process from the ranking agency to the user of higher education. Its approach to comparison has also secured wide-ranging support. The CHE methodology lacks the emotive potency of a single league table but arguably provides more data, and more useful data. It represents the most potentially transformative challenge to the present logic of global positional competition that has so far evolved, especially given its potential to become a European-wide system, and the total size of the European higher education area.

CHE surveys 130,000 students and 16,000 academic faculty in almost 250 higher education institutions. It collects data on student experiences and satisfaction, and academic recommendations on the best places in each field, focusing on 36 academic subjects each offered by a substantial number of institutions, which are updated in clusters within a three year cycle. It supplements the surveys with independent sources comprising one third of the data base. No data are taken from institutions. CHE ranks institutional departments according to each separate indicator of academic and service quality, assigning them to top third, middle third or bottom third of all institutions. The CHE refuses to integrate the indicators into a single weighted indicator either for each academic subject or for each institution, on the grounds that there is no theoretical or empirical basis for doing so. It argues that there is no 'one best university' across all areas, and 'minimal differences produced by random fluctuations may be misinterpreted as real differences' (CHE, 2006). Moreover, students have heterogeneous preferences as to mission and purpose. Accordingly, the CHE data are made available to prospective students and the public free of charge via an interactive web-enabled database. Any person can interrogate this data base by investigating and rank identified disciplines and administrative services, using their own combination of criteria (CHE, 2006), thereby creating the weightings and rankings themselves.

The CHE approach acknowledges that the definition of 'quality' is purpose-driven and open to variation and passes power over the definition from the ranker to the consumer. It also provides by far the most comprehensive set of comparative data. The data on teaching quality are limited by their subjective nature, but this is the only inter-institutional comparison on teaching executed successfully. Significantly, the CHE comparisons have proven highly successful with public and students, governments, and academic experts on rankings systems (van Dyke, 2005; Salmi & Saroyan, 2006; Usher & Savino, 2007). The CHE data collection has been extended to Switzerland, Austria, the Netherlands and Flanders, the Dutch-speaking portion of Belgium. 'The CHE ranking system is thus well positioned to develop into a European-wide system' (van der Wende, 2007b).

One difficulty with the CHE approach is that this method depends on a relatively high level of homogeneity between institutions, and between the different programs and services offered by those institutions, to enable comparison. Therefore to successfully execute the CHE approach in the United States, it would be necessary to separately compare the institutions in the different Carnegie classifications. Likewise to develop a CHE-type system for the whole of European higher education, more varied than that of Germany alone, would require a typology and separated comparisons for the different groups of institutions.

Table 3.5. Top ten European research universities as measured by publications and citations using four different indicators by the CWTS at Leiden University

	1 Raw number of research publications	*2 Citation impact (citations per publication), unmodified*	*3 Citation impact, controlled for field variation ('brute force')*	*4 Citation impact, controlled for field variation and for university size*
1	Cambridge UK	Lausanne Switz	Cambridge UK	Oxford UK
2	UC London UK	Geneva Switz	Oxford UK	Cambridge UK
3	Oxford UK	Oxford UK	UC London UK	ETH Zurich Switz.
4	Imperial C UK	Basel Switz	Imperial C UK	Lausanne Switz
5	LM Munchen Den	Karolinska Swed	Utrecht Neth	Geneva Switz
6	Paris 06 France	Erasmus Rot Neth	Helsinki Finland	Edinburgh UK
7	Milan Italy	UC London UK	ETH Zurich Switz	UC London UK
8	Utrecht Neth	Cambridge UK	Cath Leuven Belg	Erasmus Rot Neth
9	Cath Leuven Belg	Edinburgh UK	LM Munchen Den	Imperial C UK
10	Manchester UK	Zurich Switz	Karolinska Swed	Basel Switz

Source: CWTS, 2007

The Leiden metrics. The Centre for Science and Technology Studies (CWTS) at Leiden University in the Netherlands is preparing a new ranking system based solely on bibliometric indicators that it has developed (CWTS, 2007). This takes in the 400 universities with the largest number of academic publications, covering the output of at least 600 active researchers in each of those universities. The Leiden group has developed four sets of indicators:

1. Ranking by number of publications;
2. Ranking by average impact as measured by citations per publication;
3. Ranking by average impact as measured by citations per publication, modified by normalisation for academic field (controls for different rates of citation in disciplines, for example the number of citations per publication is lower in engineering and applied science than medicine). The CWTS at Leiden calls this the 'brute force' indicator;
4. Ranking by average impact as measured by citations per publication, modified by normalisation for academic field as for 3, and controlled also for institution size.

The Leiden group considers indicator 4 to be its 'crown' indicator. This measures the overall performance of a research university independent of its size, thereby removing one limitation of the SJTUIHE comparison. Other strengths of the Leiden approach compared to the Jiao Tong rankings are the modification for different rates of citation by field, removing the effect of academic field composition on the outcome of comparisons between institutions; and the avoidance of the problem of composite indicators and aggregation based on arbitrary weightings. The Leiden group has also dispensed with the Nobel indicators which are the weakest part of the SJTUIHE approach. The result is a purer measure of unit academic 'firepower' in terms of citation impact.

The Leiden group has initially applied its method to the top 100 research universities in Europe and Israel combined. The results show the sensitivity of rankings to the 'tweaking' of methodologies and raise questions about the singular nature of not just the Jiao Tong ranking but the Jiao Tong approach to publication and citation volume and citation impact. As Table 3.5 demonstrates, when the indicator is shifted from total number of publications to the quality of publications as measured by citation, the dominance of the UK universities is diminished, though the British system remains number one, and universities from Switzerland, the low countries and Scandinavia come into their own. Using the 'crown' indicator the UK and Swiss universities occupy nine of the top ten places.

A European Typology. To value diversity in comparison between institutions it must be transparent and well understood, and there must be cultural and/or administrative barriers to 'mission drift' and convergence around a single dominant model of institution, normally that of the comprehensive research university. This suggests an agreed typology of higher education institutions based on distinctive missions. The US has the Carnegie classification, which was revised in 2005 (Sapp & McCormick, 2006). In Europe a comprehensive typology is being developed at present (van Vught et al., 2005; van der Wende, 2007b). This typology is made up of a number of parallel schemes each based on a different characteristic: size of institution, legal status, type of education (degrees offered, range of disciplines), student and staff profiles, activities in research and innovation. Early work suggests there is considerable variation in the quality of potential data under the schemes. For example bibliometric indicators for basic research are well established while indicators for the socio-economic relevance of applied research are not. The areas of teaching, lifelong learning, innovation, knowledge transfer and local and regional engagement all require further work (van der Wende, 2007b). In the evolution of the typology in Europe institutions are being asked to specify their own missions, consistent with autonomy. However it is unclear whether once the typology is established institutions will be free to redefine their mission and category; and if so what could retard academic drift into higher status categories over time, which would tend to conflate categories and weaken diversity.

Further, while a typology offers the prospect of protecting diversity at the level of broad category, as the US experience shows it does not necessarily overcome the other difficulties of league tables based on holistic comparison. This confirms that to achieve useful comparative data that are consistent, transparent to the users and foster motivation in all institutions to improve performance; while sustaining diversity and avoiding the problems of partiality and validity inherent in holistic league tables; the better approach would be to use the CHE methodology of user-controlled comparison in conjunction with as typology. If this was done on the global scale, the categories of the typology could include diversity of language as well as diversity of mission and program.

The Berlin Principles. The formulation of the Berlin Principles (UNESCO/IHEP 2006), adopted on 18-20 May 2006, was led by the higher education program of UNESCO and the Institute for Higher Education Policy in Washington.

The Principles suggest that rankings and league tables should be transparent as to purpose, method and sources of data, and the weighting of indicators; use valid and relevant indicators, where possible focusing on outputs rather than inputs; use audited, verifiable and scientifically sound data; take diversity into account; and be refined and improved over time. To this extent the Principles present no challenge to the Jiao Tong approach, although those related to transparency and validity might be invoked as a basis for criticism of the Times Higher. More challenging is the requirement in the Principles that rankings 'specify the linguistic, cultural, economic, and historical contexts of the educational systems being ranked'. It is also suggested that rankings 'be compiled in such a way that eliminates or reduces errors in original data, and be organized and published in a way that errors and faults can be corrected', a nod in the direction of the CHE approach to comparison.

The formulation of the Berlin Principles was attended by much goodwill but did not meet the more fundamental criticisms that had been raised in relation to league tables in which whole institutions are represented by single numbers. Essentially the Berlin meeting went with the trend to global reputational rankings, attempting to modify that trend rather than transform it. The opportunity to encourage the spread of the CHE approach beyond Europe was missed, despite widespread support for that approach in policy circles around the world. In the Principles the problems of arbitrary selection of indicators and weightings, partial coverage and bias including language bias, disincentives to invest in local innovations with longer lead times; and the circular recycling of reputation, remain untouched.

CONCLUSION

Taken together these developments suggest that in the foreseeable future global university rankings will retain all or much of their potency, including their growing effects in relation to the movement of students and researchers, and the reshaping of national policies of research concentration and stratification. However, a more plural set of indicators and of rankings will develop; and greater specialisation in relation to mission and function, as already indicated by discipline indicators and the range of publication and citation metrics. Within Europe, where a coordinated policy-based approach to comparison and classification is in prospect, it is likely that a more nuanced and sophisticated set of indicators and machinery will be devised, in which there is less or no emphasis on holistic institutional rankings. The CHE use of consumer-selected measures is likely to be further developed. However the UK is more likely to retain its fascination with singular league tables.

Rankings function as a meta-performance indicator. They do more than 'reflect' an institution's profile and quality. The criteria used to determine each institution's position in the ranking system become meta-outputs that each institution must place on priority. In that manner rankings begin to define what 'quality' means and they shape its subsequent evolution. In the world according to the Shanghai Jiao

Tong University rankings, 'quality' means scientific research and Nobel Prizes, not teaching or community building or solving local or global problems. In the world according to the Times Higher, 'quality' means reputation, primarily; and larger staff-student ratios, research, and international staff and students; and it is partly fostered by marketing the institution across borders, which drives up Times Higher-defined performance. It seems that institutional leaders, and governments, have been given their marching orders. By shaping university and system behaviours, while standardising the definitions of outcome and output, rankings begin to decide university mission and the balance of activity. In part the power over university identity is shifted from institution itself to the ratings agencies and to the market they are helping to construct (by-passing national governments along the way!). This shift from autonomy to heteronomy in itself is very important, and carries with it serious problems for the trajectory of higher education in many countries. But these should not be overstated. The heteronomy is not complete. It scarcely touches the top institutions, which remain masters of their own identity; and the more diverse and plural are the systems of comparison and ranking, the less that any one ranking system will reshape the other institutions below the top.

Regardless, the contribution of global university rankings to market formation in the sector proceeds apace. Rankings render higher education and research more amenable to singular processes of measurement, production and distribution in the commodity form, for economic markets or quasi-market competition. Ranking of any and every university in the world, that supremely ambitious (some would say fantastical) project, for all its savage violation of diversity of mission and the specificity of values and cultures, allows us to imagine higher education and research in terms of one inter-dependent world. This in itself is a boon. Education and knowledge *have* become inter-dependent on a worldwide scale, just as economy and culture (ecology always was globally inter-dependent) and it is time this was recognised. And the possibilities are open. But a rankings mediated market, based on positional value, is too closed. The problem posed by rankings of the holistic league table kind is that, rather than presenting the world to itself, allowing all the complexity, variety and unevenness to emerge, rankings grossly simplify and reify the complexity. Institutions from diverse traditions are measured against the nations that have dominated the world for the last 200 years. In this setting, global unevenness shows itself not as poverty of educational means, which would thereby become a joint global responsibility; but poverty of educational performance. The latter is readily blamed on the victim, in the manner of markets.

This suggests that in developing the global comparisons of universities and of national systems, more is needed than just the development of a wider range of indicators of performance, and formation of choice based systems of comparison. After all, the objective is not to rank every university as an end in itself (is it?), or for one's own country to win the global contest (is it?), as if higher education was the World Cup. It is to everywhere lift the contribution of higher education and research to public and private goods. Cross-border comparisons should map and model the *capacity* of institutions and nations in education and research, and their various contributions to private and public goods: local, national and global.

REFERENCES

Altbach, P., (2006). The dilemmas of ranking. *International Higher Education*, 42.
Anderson, B. (1983). *Imagined communities: Reflections on the origin and spread of nationalism*. London: Verso.
Bourdieu,. P. (1984). *Distinction: A social critique of the judgment of taste*. R. Nice (transl.). London: Routledge.
Centre for Higher Education Development, CHE (2006). Study and Research in Germany. University rankings, published in association with Die Zeit. Accessed on 16 December 2006 at: http://www.daad.de/deutschland/studium/hochschulranking/04708.en.html
Centre for Science and Technology Studies, Leiden University, (CWTS) (2007). The Ledien ranking. Accessed 20 June 2007 at: http://www.cwts.nl/cwts/LeidenRankingWebSite.html
Clark, B. (1998). *Creating entrepreneurial universities: Organizational pathways of transformation*. Oxford: Pergamon Press.
Dill, D. & Soo, M. (2005). Academic quality, league tables, and public policy: A cross-national analysis of university rankings. *Higher Education*, 49, 495-533.
Dyke, N. van (2005). Twenty years of university report cards. *Higher Educatiton in Europe*, 30(2), 103-125
Foucault, M. (1972). *The archaeology of knowledge*. A.M. Sheridan Smith (transl.). London: Tavistock.
Frank, R. (2001). Higher education: The ultimate winner-take-all market? In M. Devlin & J. Meyerson (Eds.) *Forum futures: Exploring the future of higher education*. Forum Strategy Series Vol. 3. Jossey-Bass: San Francisco.
Frank, R. & Cook, P. (1995). *The winner-take-all society*. New York: The Free Press.
Henkel, M. (2005). Academic identity and autonomy in a changing policy environment. *Higher Education*, 49, 155-176.
Henkel, M. (2007). Can academic autonomy survive in the knowledge society? A perspective from Britain. *Higher Education Research and Development*, 26(1), 87-99.
Hirsch, F. (1976). *Social limits to growth*. Cambridge: Harvard University Press.
Institute for Scientific Information, ISI-Thomson (2006). Data on highly cited researchers, ISIHighlyCited.com. Accessed 15 August 2006 att: http://isihighlycited.com/
Kirp, D. (2004). *Shakespeare, Einstein and the bottom-line: The marketing of higher education*. Cambridge MA : Harvard University Press.
Liu, N. & Cheng, Y. (2005). The academic ranking of world universities. *Higher Education in Europe*, 30(2), 127-136
Liu, N. & Liu, L. (2005). University rankings in China. *Higher Education in Europe*, 30(2), 217-227.
Linguasphere Observatory (2006). *Linguasphere table of the world's major spoken languages 1999-2000*. Accessed on 2 April 2006 at http://www.linguasphere.org/language.html
Marginson, S. & Considine, M. (2000). *The enterprise university: Power, governance and reinvention in Australia*. Cambridge: Cambridge University Press.
Marginson, S. and Mollis, M. (2001). 'The door opens and the tiger leaps': Theories and reflexivities of comparative education for a global millennium. *Comparative Education Review*, 45(4), 581-615.
Marginson, S. & van der Wende, M. (2007). *Globalization and Higher Education*. Paper prepared for Organisation for Economic Cooperation and Development. Paris: OECD. Accessed 31 May 2007 at: http://www.cshe.unimelb.edu.au/people/staff_pages/Marginson/Marginson&van_der_Wende2006O ECD.pdf
McCormick, A. C. & C. Zhao (2005). Rethinking and reframing the Carnegie Classification. *Change* 37(5): 51-57.
Musselin, C. (2005). European academic labour markets in transition. *Higher Education*, 49, 135-154.
Newman, J. (1996/1899). *The idea of a university*. F. Turner (Ed.). Yale University Press: New Haven.
Nowotny, H., Scott, P., & Gibbons, M. (2001). *Rethinking science: Knowledge and the public in an age of uncertainty*. Cambridge: Polity.

Ortmann, A. (2002) *Capital romance: why Wall Street fell in love with higher education.* The For-Profit Higher Education Research Project at the Curry School of Educaiton, University of Virginia. Accessed on 11 April 2006 at: http://curry.edschool.virginia.edu/forprofit/workingpapers.htm

Rhoads, R. & Torres, C. (2006). *The university, state, and market: The political economy of globalization in the Americas.* Stanford: Stanford University Press.

Salmi, J. & Saroyan, A. (2006). *League tables as policy instruments: Uses and misuses.* OECD/IMHE & Hochschulrektorenkonferenz, Germany, Workshop 4-5 December, Bonn. See Organization for Economic Cooperation and Development, Institutional Management in Higher Education program, Accessed 10 March 2007 at: http://www.oecd.org

Sapp, M. & McCormick, A. (2006). Revision of the Carnegie classifications. Special Issue, AIR Alert 25, Update 1.

Shanghai Jiao Tong University Institute of Higher Education, SJTUIHE (2007). *Academic ranking of world universities.* Accessed 20 June 2007 at: http://ed.sjtu.edu.cn/ranking.htm

Sowter, B. (2007). *THES-QS world university rankings.* Symposium on International Trends in University Rankings and Classifications. Griffith University, Brisbane, 12 February. Accessed on 19 June 2007 at: http://www.griffith.edu.au/conference/university-rankings/

The Times Higher (2006). World university rankings. *The Times Higher Education Supplement.* Accessed on 30 November 2006 at: www.thes.co.uk [subscription required]

Usher, A. & Savino, M. (2006). *A world of difference: A global survey of university league tables.* Accessed on 2 April 2006 at: http://www.educationalpolicy.org

UNESCO/IHEP (2006) *International partnership issues groundbreaking principles on ranking of higher education institutions.* Accessed on 3 June 2006 at: http://www.che.de/downloads/Berlin_Principles_IREG_534.pdf

U.S. News and World Report (2006). *America's best colleges, 2007 edition.* Washington: USNWR.

Valimaa, J. (2004). Nationalisation, localization and globalization in Finnish higher education. *Higher Education,* 48, 27-54.

Valimaa, J. (2005). Globalization in the concept of Nordic Higher Education. In A. Arimoto, F. Huang, and K. Yokoyama (Eds.). *Globalization and higher education.* International Publications Series 9, Research Institute for Higher Education, Hiroshima University. Accessed 10 February 2006 at: http://en.rihe.hiroshima-u.ac.jp/pl_default_2.php?bid=63653

Van der Wende (2007a). European responses to global competition in higher education. Symposium 'Crisis of the publics', Centre for Studies in Higher Education, UC Berkeley, 26-27 March. Accessed 1 June 2007 at: http://cshe.berkeley.edu/events/crisis/docs/ [password required]

Van der Wende (2007b). Rankings and classifications in higher education: A European perspective. 2nd Leiden University International Symposium on Ranking, 2 & 3 February.

Williams, R. & Van Dyke, N. (2006). *Rating major disciplines in Australian universities: Perceptions and reality.* Melbourne: Melbourne Institute of Applied Economic and Social Research.

World Bank (2006) World Bank data and statistics. Accessed 15 August 2006 at: http://web.worldbank.org/WBSITE/EXTERNAL/DATASTATISTICS/0,,contentMDK:20399244~menuPK:1192694~pagePK:64133150~piPK:64133175~theSitePK:239419,00.html

Simon Marginson
Centre for the Study of Higher Education
University of Melbourne
Melbourne, Australia

LEONARD J. WAKS

4. IN THE SHADOW OF THE RUINS:

Globalization and the Rise of Corporate Universities

INTRODUCTION

The term 'corporate university', while introduced as early as the 1970s, has come into wide use only in the last decade. It is used in several vague and shifting ways, and at this moment is applied to at least three distinct kinds of organizations as well as hybrids combining them.

Three Types of Corporate Universities

Corporate universities of the first type are established, mainstream, non-profit universities, public or private, which are adapting to economic and political pressures or technological opportunities by adapting managerial practices of modern for-profit corporations. These universities are 'corporate' in the sense that they behave increasingly like large for-profit corporate firms. Established universities in the United States, Britain, Canada, Australia, are now 'corporate' in this sense, according to many scholars. Some encourage this trend (Clark,1998). Others explain it in historical or sociological terms (Readings, 1996), carefully document it (Slaughter and Leslie, 1997), differentiate trends in specific regions (Currie, et. al., 2003), or simply decry it (Aronowitz, 2000).

Second, the term is applied to newly established, highly innovative universities that operate as for-profit corporations, but satisfy the political and legal requirements for university status, and meet the standards of accrediting bodies. Among the best known of these universities are the University of Phoenix, DeVry, Strayer, and Jones International. These are 'corporate' in the straightforward sense that they are large for-profit corporations.

Finally the term is used to refer to new organizations operating within, and providing education and training services for, for-profit corporate firms. These organizations have typically replaced training departments previously existing within the human resources divisions of these firms. Today they are located closer to the office of the chief executive, and are focused entirely upon the strategic missions of the firm. They disseminate knowledge and information to personnel of the firm and its suppliers and customers. But they also 'harvest' new knowledge generated by these workers, circulate it throughout the firm's network of suppliers and customers for 'knowledge re-use,' and 'capture' it as intellectual property of the firm. This complex of activities is known as 'knowledge management'.

McDonald's Hamburger University was an early, and much ridiculed, corporate university. Today several hundred corporate universities exist in the United States,

Simon Marginson (ed.), Prospects of Higher Education, 101–119.

and especially since 1998 they have spread to European and Asia-Pacific corporations Dickson, 2002; Arnauld-Coulon, 2002). Motorola University and Marriott University are prominent examples. A few of these new meet political and legal requirements for university status, and grant their own diplomas and graduate degrees (Allen, 2002). More often, they provide non-degree certificates, and partner with non-profit or for-profit universities (corporate universities of the first or second types) to offer diploma and graduate degree programs. Like corporate logos, these universities are 'corporate' in that they are an asset of the corporations -- they belong to them and exist to serve their goals.

Corporate Universities in the Shadow of the Ruins

The knowledgeable reader will not have missed the oblique reference in my title to the late Bill Reading's remarkable essay *The university in ruins* (1996). The title pays homage to Readings, and in particular to his account of established universities as organizations that have lost their received social role within the era of globalization.

For Readings the modern university has been an agency of the modern nation state. National cultures emerged in the modern period as carriers of the symbol systems, beliefs and values of their respective states. The modern universities had the mission of creating and inculcating a national culture through both research and teaching. The notion of 'culture' combined two senses: culture as the sum of knowledge, and culture as cultivation of character. The university, combining research and teaching, was responsible for both the production and inculcation of national self-knowledge. In the research university, scholarship and teaching functions converged as students acquired the latest, most up-to-date, version of the national knowledge culture from its leading exponents in the very process of working alongside them to create it.

But in the era of globalization the function of the nation state has been transformed. As Readings states, 'the university no longer has to safeguard and propagate a national culture because the nation-state is no longer the major site at which capital reproduces itself' (Readings, 1996, p. 13). The university is no longer called upon to train national citizens and leaders, because the cosmopolitan world citizen of global society is attached to the modern corporate order more securely than to the nation state. As a result, the idea of a national culture can no longer provide a secure and common external reference point for either knowledge production or cultivation of character in the contemporary university. The established university is thus in 'ruins' - it has 'outlived itself' (Readings, 1996, p.6). Buildings stand, students attend, scholars produce knowledge, diplomas are awarded. But the knowledge produced, and the character cultivated, do not cohere around a national spiritual centre. Universities are now struggling to find a new identity as relatively autonomous corporations, marketing their wares globally to individual and corporate customers. They are thus evolving into 'corporate universities' of the first type.

I accept Readings' analysis as a starting point, but with a significant qualification: the transformation of the university from a centre of national culture to a predominantly vocational training school began as early as the post-world war 2 period when the university first became a mass institution. Indeed, the origins of the modern university are found in the expanding need for personnel in bureaucratic state agencies and industrial firms in the nineteenth century. By then elite families could no longer supply personnel in sufficient numbers, or with skills and attitudes appropriate to modern conditions. The universities recruited and trained middle class youths for these bureaucracies. The national cultures embodied the symbol systems within which discourses of leadership were conducted, and university graduates were initiated into these discourses.

The economies of neo-liberal, pecuniary societies, however, forged ever-expanding occupational opportunities for those with the knowledge and skills conveyed in universities. Neo-liberal states encouraged expansion of the universities for larger and larger fractions of the population, in order to maintain administrative control of young people, and manipulate their economic incentives. Educational qualifications for occupations of every sort were inflated. A much larger fraction of the population gained access to the university, and needed a diploma to qualify for respectable employment.

The graduates of mass universities were 'educated personnel' but no longer, in the general case, leaders involved in discourses about the future direction of society. As such, the symbol systems in which these discourses were conducted were akin to foreign languages. Universities continued to provide at least an attenuated version of the liberal-modern curriculum of the nineteenth century university, with the liberal component involving required course work in national history and literature. But occupationally oriented students from working families lacking prior academic experience, including older 'non-traditional' students and those from disadvantaged minority groups, increasingly found such courses a burden and an unwarranted imposition. The 'culture' conveyed was detached from their collective memories or aspirations. Some of these students demanded an expanded 'multi-cultural' humanities curriculum, but most wanted merely to pursue a diploma certifying them as occupationally qualified. To the despair of the faculty, they wanted to be 'educated' in a sense that involved no 'cultural' encumbrance whatsoever.

The mass universities were thus on the way to becoming vocational schools even before the recent transformation of the nation state. But in the wake of globalization, as Readings has argued, even elite universities training the leadership class are struggling to re-invent themselves. These universities remain positioned at the centre of higher education but the meanings generating and sustaining that position are evaporating.

Power is shifting from national centres to global networks. There is no coherent global institutional order serving as the centre around which a global culture, global ideology, or global leadership can form, and hence no coherent 'philosophy' to regulate knowledge production or teaching for a leadership class. The universities are seeking a new definition as providers of knowledge and attitude 'products' suitable for the economic and technical arenas of global society. But as their leaders struggle internally to re-form themselves, against the resistance of their brigades, they are increasingly encountering corporate competitors who, while lacking their legitimacy within higher education, possess objectives, internal structures, and personnel adapted expressly for skirmishes in that market terrain.

Understanding the Corporate University

The goal of this chapter is to situate the corporate university in a broader institutional account of contemporary education. I group corporate universities, along with other educational organizations coming to the fore in the era of global markets and networks, as 'shadow organizations'. These other organizations include private firms supplying educational services to public school authorities (for example Westinghouse, Berlitz); tax supported but weakly regulated charter schools operated by private non-profit or for-profit organizations (for example Chris Whittle's Edison Corporation); technical academies sponsored and controlled by corporations (for example Microsoft, Dell and Cisco Systems) offering curricular programs within mainstream schools; and even home schools and home-school support organizations, as well as various hybrids of such forms. Thinking of these as 'shadow organizations' suggests a way of thinking and talking about them.

I account for these organizations as ordained by an institutional fragment breaking off from the central educational institution of post-industrial societies, as will be discussed in the next part of the chapter. I label such fragments 'shadow institutions,' the organizations they ordain 'shadow organizations,' and the positions and roles within these organizations 'shadow' positions. While the word 'shadow' is usually used as either a noun or verb, and these uses are highly suggestive in the study of shadow institutions, there is also a recognized adjectival use, in which, according to the American Heritage Dictionary (4th edition., 2000) 'a shadow -- ' is 'a -- not having official status'. It is in this sense that there are shadow cabinets, shadow governments of exiled leaders, and shadow economies. The idea of an 'official status' is not sufficiently clear, so I will construct a more definite conception of shadow institutions suitable to my larger interpretive purpose in the middle section of the paper. I close by applying this account to actual corporate universities, showing that it provides a fitting and suggestive way of thinking and talking about these organizations, in the final section.

INSTITUTIONAL FRAGMENTATION

Organizations and Institutions

The root word in 'organization' is 'organ,' derived from the Greek 'erg' meaning work. In the relevant sense, an organ is a differentiated part of an organizm, adapted for a specific job or function. The Oxford English Dictionary makes this sense more specific: an organ is a part or member of an animal or plant body adapted by its structure for a particular vital function, as digestion, respiration, perception or locomotion. Something is 'organic' when it is composed of distinct but structurally united organs, which serve distinct purposes but are coordinated for the united action of the whole. Thus the key notions contained in the concept of 'organization' are (i) work: tasks or vital functions, (ii) differentiation: of multiple parts or 'organs' performing these tasks, and (iii) structural unity: parts arranged to subserve a vitality of the whole. The biological roots are central.

The term 'institution' has many over-lapping senses, all related to 'organization' in one way or another. In one sense we speak of 'social institutions' such as the educational, political, or economic institutions of society. In this sense the economic institution certainly has much to do with e.g., business firms, but is not itself a firm or any other organization. But we also do speak of firms as institutions, for example an insurance company, as an established institution. In this sense, an institution is an organization. The same senses appear to apply to educational, political and cultural institutions. We speak broadly of the cultural institution of The United States. But we also speak, in another sense, of the various churches, universities, museums, and similar organizations as cultural institutions. For example the Smithsonian Institution is a cultural organization operating museums.

It is in the elusive first sense ('social institutions') that we may distinguish institutions and institutional change from organizations and organizational change. We can begin to understand this sense by starting with the verb 'to institute', which derives from the Latin instituere, to set up, establish. The verb has the sense of to begin, to introduce, to bring into use or practice, to found. Institution, as the act of instituting, has the sense of establishing, ordaining, or setting in motion, of giving form or order to a thing. To ordain, for its part, in its earliest fourteenth century use, means to arrange, put in order, make ready, prepare, to furnish, fit out, equip. These roots can be observed in the standard sociological sense of an institution. For Hamilton (1932), social institutions are the ways of thought and action of some prevalence or permanence, embedded in the habits of a group or the customs of a people. They fix the confines and impose form upon the activities of human beings. Hamilton emphasizes the ubiquity of social institutions: 'about every urge of mankind an institution grows up; the expression of every taste and capacity is crowded into an institutional mold.' Eisenstadt's (1972) account is only slightly weaker. Institutions are 'regulative principles which organize most of the activities of individuals into definite organizational patterns from the point of view of some of the perennial, basic problems of any society or ordered social life' (p. 410).

Following these accounts, by institutions I will mean the social arrangements establishing, ordaining or authorizing, the ideas, norms, organizations, and frameworks regulating human interaction in the primary areas of human life. The core institutional systems include family and kinship, politics, economics, education, and religion and culture, and the system of social stratification.

Institutional spheres overlap and connect. Occupational education, for example, is mandated and funded by the political institution, is located in the educational institution, and is preparatory for participation in the economic institution. It consists in large measure in learning the use of various symbols (the cultural institution), and is distributed according to the norms of stratification. These linkages of institutional spheres bind the various activities of social life together, and constitute the institutional order as a social whole or 'a society'.

Institutions build up cumulatively. Each holds within itself the survivals of behavioural and symbolic forms from the past, 'vestiges of design and accident, the stuff of idea and custom, from many ages, societies, civilizations, or climates of opinion' (Hamilton, 1932, p. 88). Institutions form the background of organizational life, shaping the habits, ideas, and norms regulating the existing organizations which they authorize or ordain, and the roles of those occupying the positions within them. They make up the cognitive and affective surround of organizations and organizational behaviour. Goals and practices of organizations must conform to institutional norms and ideas to retain their social position.

Institutions, however, are not uniform. They may fix confines and regulate forms of behaviour, but the areas confined are not homogeneous. A diversity of beliefs and values and practices exists. A predominant consensus at the institutional centre weakens as one approaches the margins. Social actors with new or simply different ideas, values, and practices can maintain positions in organizations ordained by the institution. They may be considered 'fringe' actors in mainstream organizations, and they may even form 'fringe' organizations operating within institutional confines. But the confines remain, and if these actors go 'beyond the fringe' institutional authorization will be withdrawn.

Institutions are also not static, they are pressured to change functionally or structurally as the institutions surrounding them change. Thus beliefs, values and norms change, new practices evolve and struggle for legitimacy, the confines weaken, organizational forms evolve. Older, mainstream organizations may become peripheral or disappear. New organizations may emerge and after a time gain legitimacy. Fringe practices may become first acceptable, then commonplace.

Institutional and Organizational Change

Institutional arrangements, like the specific organizations they authorize, change over time. Institutions are adaptive tools for meeting basic human purposes. They are connected and coordinated components of a society's life. As one sphere changes others must adapt to maintain stability. Until then people will be increasingly frustrated in attempts to meet everyday needs and achieve everyday purposes. As people then re-adjust their ideas and habits in response to this frustration, institutions eo ipso begin to change. 'As one social system passes into another and the values and manner of living are transformed, one institution gives way to another better adapted to the times' (Hamilton, 1932, p. 85). People gradually withdraw their authorization of established organizations and commit to new organizations promising better intra- and inter-institutional re-alignment.

As institutions are deeply rooted in history and tradition, in 'the stuff of idea and custom,' as inter-institutional linkages form and solidify, and as institutional actors are attached to the economy by virtue of their established positions, institutions rarely change in simple linear fashion. Instead, the change process is marked by contradiction and fragmentation, and is punctuated by reaction and resistance, rebellion and relinquishment. When factors in the institutional environment signal the need for change, typically established organizations initially resist. They may reject, or contest and counter, the signals, and cling tightly to their established forms and functions. Their leaders may make stirring appeals to history and tradition. When influential thought leaders in society, positioned outside a specific institutional area, call out for organizational change and circulate their ideas in the media, they are likely to be countered by complex arguments by internal thought leaders that may persuade insiders with vested interests more than outsiders. Even when edicts for change are issued by policy makers, they are deflected, diluted, or channelled into insignificant and short-lived programmatic efforts at the organizational periphery. Internal leaders buffer external demands and work with insiders to fashion a fit between external demands and long- established practices.

Even more important, older institutional practices are shored up by governmental and other forms of authority, and institutional categories entrenched deeply in the public mind. In the case of the university, these practices may be inscribed in government licensing regulations for higher education, the standards used by accreditation agencies, and the categories used to discuss higher education as a topic of public concern. As a result, organizations adopting new practices that respond to new needs may become institutionally vulnerable. They may lose their licences or accreditation or public recognition as 'true' institutions of higher education. In this way the institutional centre can hold long after the external factors that put it in this place have passed.

Institutional Fragmentation

But gradually the beliefs and values of sub-groups in society begin to change. The broader social institution (by definition) begins to withdraw authorization from established organizations in their existing forms. Contradictions in the institution become sharper and problems caused by established, hide-bound organizations become deeper. The normative outlines of reformed organizations become clearer.

At this point the institution begins to fragment. Most members of society remain attached to the old organizations and cling to the old practices, though many will be going through the motions to a greater or lesser extent. But external pressures for change can no longer be ignored. Organizational leaders begin to initiate reforms despite resistance. Meanwhile cutting-edge fractions of society come entirely to reject the older organizations. New beliefs and values circulate, along with more concrete proposals for new organizations and practices. The cognitive and affective components of a new institutional formation are now in place.

Eventually 'organizational entrepreneurs,' within and beyond the established organizations, may attempt to embody the new ideas in concrete organizational forms. Early adopters can then withdraw not merely their commitments, but their participation from the established organizations to take part in the new ones.

At this point the institutional status of the new organizations may remain indeterminate. While in conformity with emerging beliefs and values and models of rational action, and thus acceptable to risk-taking early adopters, they still look strange – absurd, bizarre, crazy, weird -- to a great many people. An established legal status, political support, accreditation, or viable economic existence may still be lacking. A prolonged process of institutional give and take, of change and adjustment may take place. In both established organizations and new ones, new practices and new ways of talking about them may be needed before the new organizations can be assigned a secure place in a new institutional order, alongside the established but now reformed older organizations.

SHADOW INSTITUTIONS AND ORGANIZATIONS

The institutional fragments – the clusters of beliefs and values, norms, organizational images and proposed practices - taking shape at the periphery of an established social institution, and generating leading to new organizational forms, I will call 'shadow institutions.' The organizations brought into existence in this shadowy institutional space I shall call 'shadow organizations' and the positions within these organizations 'shadow positions'.

Shadow Institutions, Official Status, and Social Authority

'Shadow' has a recognized adjectival use, defined as 'not having official status'. In this sense the term is applied to organizations peripheral to core institutional systems, as in 'shadow government' and 'shadow economy'. The notion of 'having an official status' is not entirely clear, however, and will require a closer analysis. 'Official' means, in the first instance, 'of or pertaining to office.' 'Office' in turn is derived from the Latin opus (work) + facere (to make or to do). An office is in the primary sense 'a special duty, charge, or position conferred by an exercise of governmental authority and for a public purpose' (Oxford English Dictionary). An 'official' (noun) is a person with such an office. These usages imply a definition of 'shadow institutions' (and the organizations and positions they ordain or authorize) in terms of their lack of authority as conferred by an exercise of government. This government-related notion fits some adjectival uses of 'shadow' well. The 'shadow economy' may be defined in terms of economic activity not regulated by licenses, qualifications, tax codes, and so on, administered by the state.

Still, such a definition appears overly restrictive, in that it suggests an essential, and even determinative, role for the state in authorizing rules and practices of social life. Some institutional formations lacking specifically governmental authorization may nonetheless be established institutions capable of possessing their own shadows. The shadow government of Britain is not the official government. But does it have an 'official status'? While its shadow ministers lack official responsibility for state agencies, multi-party democracy and the opposition in Parliament are institutionalized and serve a recognized public purpose, the development of alternative lines of public policy to be considered by the electorate. By contrast there are 'shadow governments' composed for example of leaders of exile groups or suppressed internal majorities, whose primary aim is to dislodge an official government they regard as illegitimate. These shadow organizations may also have an official status, the status of illegal organizations.

While many modern nation-states including Britain continue to have an established church, they recognize other established religions as elements of the mainstream religious institution. And in the United States, where the constitution prohibits government entanglement in religion, there is nonetheless an authorized religious establishment and a well-defined central religious institution. Authorization of institutions does not require conferring governmental status. This does not imply that all religious organizations are on an equal footing. Many mainstream churches have fringe elements (for example Roman Catholic anti-abortion terrorists), and many religious groups organize 'beyond the fringe': consider James Jones' Jonestown Community which committed mass suicide in Guyana in 1978, and David Koresh's Branch Davidians who perished after a fifty-one day stand-off with federal authorities in Waco, Texas in 1993. These can be conceived as 'shadow religions' ordaining their own churches and ministers. It is arbitrary to consider mainstream American religion a 'shadow institution' simply because it lacks official status conferred by an exercise of governmental authority. Instead, the sources of social authority are popular assent and conformity, and linkage to other institutions within the institutional order. The government is only one of these institutions, and not an essentially privileged one. Conferral of official status by governmental authority is certainly an important element in making shadow institutions mainstream, but it is not always either necessary or sufficient.

Institutional spaces, as discussed above, are heterogeneous. Major institutions may have at a given time a central zone, fringe elements maintaining social authority through connection to the mainstream, and shadow elements fragmenting off from it. In addition to shadow governments and economies, we should expect to have shadow churches, families, cultural and professional institutions, and schools – for example, religious cults, gay marriages, off-off-Broadway theatres, herbalist doctors, cyber-charter schools - and also corporate universities.

The term 'shadow institution' is suggestive because of the many connotations it borrows from the word 'shadow'. In its primary sense a shadow is an area either not illuminated (the umbra) or only partially illuminated (the penumbra) because of the interception of radiation by an opaque object between the area and the source of illumination. In a related but derivative sense we speak of a shadow cast not by a light, but by a dominating monument, region, presence or influence. Hence 'in the shadow of --- ', for example a scientist that works in the shadow of his lab director.

The notion of 'shadow institution' suggested here is that of an institutional fragment obscured by the attention commanded by organizations at the institutional centre. Corporate universities are new formations that remain obscured by established universities. They exist in the shadows of these organizations. A history of higher education in the 1970s, for example, would almost certainly neglect McDonald's Hamburger University. Even today, scholarship on higher education routinely neglects both for-profit adult convenience university corporations such as DeVry and the University of Phoenix and universities run as in-house education and knowledge management arms of corporations such as Motorola or Marriott

Universities. From the perspective of higher education as a field of study, these organizations exist, like Asian sweatshops or sex slave brothels, in a shadowy world most scholars may have heard about but never even wished to encounter.

In another sense a shadow is an 'imperfect imitation or copy' and in still another one 'a mirrored image or reflection'. Here the idea of a shadow institution suggests one that copies, mirrors, or reflects an original institutional model, but imperfectly. In a related sense a shadow is an unreal appearance, a delusive semblance or image. Organizations in the shadow area of higher education are often conceived as imperfect models of, or delusive semblances of, established universities. While called 'universities' they lack serious libraries or laboratories; and their teaching is shaped by the profit motive rather than the search for truth. To call the University of Phoenix, for example, a 'shadow university' in this sense would be to presuppose a normative status for established universities. It is in this sense that Stanley Aronowitz contrasts corporate universities with 'true higher education.'

In yet another sense a shadow is 'a faint or obscure indication or fore-shadowing,.' or 'an omen or pre-figuration'. Here a dominant future form casts its shadow back into the past. In this sense the 'shadow institution' is not yet firmly established. Its constituent beliefs, values, norms, organizations and practices are still 'faint': dim, blurry, indefinite, but providing a hazy image of a future institutional form that looms large. McDonald's University, or the University of Phoenix might be said to fore-shadow the university of the future, as California fore-shadows social and cultural developments throughout the United States.

I will now sharpen my definition: shadow institutions are (a) institutional fragments breaking off from central or predominant institutional formations that (b) exist in the shadows of these and are obscured by them, in that they are largely disregarded in authoritative discourses about them. When they are (c) regarded from the mainstream vantage point, shadow organizations are either invisible or appear as imperfect, and even delusive, copies or imitations of the 'real' thing. But (d) from the vantage point of their institutional entrepreneurs and their followers, shadow institutions, by responding to inescapable external pressures resisted by mainstream actors, foreshadow the future. They are omens of things to come.

CORPORATE UNIVERSITIES AS SHADOW INSTITUTIONS

This final section considers the adult university corporation and the university division of non-educational corporations as shadow institutions.

Type Two Corporate Universities -- Adult University Corporations

Adult convenience universities formed as a merger of the proprietary occupational education and the degree granting higher education sectors, beginning in the 1970s. This had several causes. One was post-industrialization, the emergence of the knowledge economy and the resulting expansion of adult education. A second was the parallel emergence of new information and communication technologies, transforming the workplace and opening new possibilities for delivery of education and training. A third was the increased attention to managerial issues in higher education due to declining 'traditional' markets and cost over-runs (Keller, 2000).

Adult universities as institutional fragments. Most of the 'non-traditional' adult students in mass universities were enrolled to acquire new skills and qualifications to maintain employability in the post-industrial economy. A large proportion consisted of full time workers drawn from working class families. Many were single parents drawn from ethnic minority groups (Ruch, 2001, p. 31). Mainstream universities were slow to respond to their non-traditional needs and expectations. These included convenient hours and locations, narrowly focused occupational programs, and suitable curricular and instructional methods. They were on the one hand, older and more responsible than traditional students. On the other had they were less academically sophisticated and more in need of academic support services. These students occupied an unserved market niche, and for-profit adult convenience universities sought to occupy it. I will focus on the best known of these, the University of Phoenix.

In the late 1970s John Sperling, a scholar with impeccable credentials who was teaching humanities at San Jose State University was awarded a grant to facilitate a group of public employees in obtaining academic degrees: masters degrees for teachers with bachelors, and bachelors for policemen with associates degrees. Sperling had won a previous grant for developing an economics course for high school seniors, in the course of which he developed a standardized teaching-learning system that he felt could be applied successfully to non-traditional college students. The system involved a standard course manual with specific learning objectives, offered to small groups led by teachers committed to the specifics of the manual. The adult program would be offered one course at a time, over five or six weeks, rather than as one among several courses taken in a standard academic term. Students would be granted advanced standing for relevant life experience. This system violated academic values, for example teachers' rights to shape their courses, and was at odds with standard academic practices. Finding himself unable to navigate the academic obstacles at San Jose State, Sperling formed an educational corporation and offered to provide his adult program, as an external contractor, to the University of San Francisco, a Jesuit college struggling with a budget crisis. The University's leaders, tempted by this new market and the overhead on the grant, were enthusiastic, and cut some internal red tape to get the program accepted. But it ran up against faculty committees, the state licensing authority, and the regional accreditation body - the exoskeleton confining academic practice. Eventually the accreditation agency, by threatening to withdraw accreditation, was able to force Sperling and the academic leaders supporting him out of the University of San Francisco. Sperling found himself 'beyond the fringe'.

As a result, Sperling, the self-described 'unintentional entrepreneur, was compelled to create his own degree-granting university. He moved to Arizona, which did not regulate universities by licence laws and had a more sympathetic regional accreditation board. Eventually he established the University of Phoenix. From the start he conceived his venture as a 'cultural battle' (Sperling, 2000, p. 79) between the existing higher educational culture and an 'idea' (p. 91). He became an 'institutional entrepreneur' shaping a new institutional fragment broken off from established higher education. He no longer represented merely a threat to market share but a 'radical change' in institutional culture (Ruch, 2001, p. 47).

Adult universities as imperfect or delusive copies – 'diploma mills'. Although operating in the shadows of mainstream universities, the University of Phoenix quickly came within the sights of the university establishment. Sperling met with 'hostility bordering on rage' from faculty members and university officials. His for-profit structure and standardized operating methods were the 'ultimate apostasy' (Sperling, 2000, p. 2). His university was condemned by university officials as a 'diploma mill'. The Arizona Regents, the governing arm of the three state universities, fought against his accreditation. They introduced 'anti-diploma mill' legislation and threatened to withdrawn from the regional accreditation body. Phoenix has had to fight costly and acrimonious legislative and accreditation battles wherever it has operated. Established universities have refused to accept its credits for transfer into their programs.

Adult university corporations obscured by established universities. Although Sperling and other institutional entrepreneurs in higher education established for-profit corporations in the late 1970s and 1980s. they laboured in 'obscurity' (Sperling, 2000, p. 158), and as 'invisible partners' (Ruch, 2001, p. 60) in the shadows of non-profit universities until the mid 1990s. Even in 1992, when Phoenix began a major expansion program with the goal of being '20 minute commute from 65 per cent of the urban population of the United States', its public invisibility was a major obstacle. It invested heavily in lobbying and publicity to make itself known to governments and public (Sperling, 2000, p. 178).

Adult universities fore-shadowing the future. However since the mid 1990s, for-profit universities operating globally as public corporations have achieved broad public and political acceptance and accreditation. They have experienced rapid enrolment and financial growth. Major players now include The Apollo Group (University of Phoenix and other educational divisions), Argosy Education Group (American Schools of Professional Psychology), Education Management Corporation (Art Institutes International), DeVry Inc., (DeVry Institutes of Technology) and Strayer Education Inc. (Strayer University).

In 2000 they operated from almost 250 sites and over 100 campuses, serving more than 200,000 students. Most of the growth has been since 1996. While most non-profit universities, public and private, faced budgetary crises in 2003, the for-proft sector posted record gains especially in their online divisions. Growth in the online enrolment of the University of Phoenix was 81 per cent in 2001, 70 per cent in 2002 and 61 per cent in 2003. From 2002 to 2003 DeVry's online operations grew from 3,800 to 9,000 students. Sylvan Inc., known for their Sylvan Learning Centres offering tutoring services to K-12 students, has more than 100,000 students in Adult University campuses (Farrell, 2004)). They recently sold the K-12 tutoring business division to concentrate on adult universities, operating online and face-to-face in the United States, Chile, France Mexico, Spain, Switzerland and India.

Sperling's initial efforts have indeed fore-shadowed the direction of established higher education organizations. In the neo-liberal climate of the current era, the University of Phoenix has been welcomed 'with open arms' by states that had previously rejected it out of hand, and accreditation agencies have accommodated its approaches (Sperling, 2000, p. 181). It is now accepted by many as 'the very model of a twenty-first century university.' Though Phoenix is still compelled to fight against the 'rearguard action of educational bureaucrats doggedly protecting the market share of their local institutions' (p. 80), mainstream universities have adopted its corporate practices, though they still deny it (Sperling, p. 91).

Type Three Corporate Universities -- Units of Non-Educational Corporations

Corporate universities of the third type, university divisions within for-profit non-educational organizations focused on the strategic missions of those organizations, grew out of dissatisfaction with standard business school education and the MBA, experienced in the 1970s and 1980s by both corporate managers and innovative business school faculty members. Unlike adult convenience universities, these schools aimed at top rather than bottom and middle level employees. Many top managers found traditional MBAs to be technically proficient but lacking in the 'soft skills' increasingly needed for global competition, including communications across disciplines, functional units, and employee levels; flexible and creative problem solving; and working cooperatively in teams (Allen, 2002, p. 5). Top managers also wanted the training of their younger counterparts to be aligned with the mission and markets of the firm. As managerial training had moved into the universities it had been redefined away from managerial practice, becoming increasingly general, theoretical, and in a word, academic. This was inevitable, as professors lacked intimate knowledge of specific firms and industries.

Many firms moved their managerial training functions in-house. While they already had human resources departments, many top managers saw these as suffering from the same problem as universities, isolation from the firms' core operations (Barley, 2002, p. 43). Thus corporate universities were born. Firms developed education and training units answerable directly to top management, and closely linked to the core operations of the firm.

Corporate universities have responded to new technological, organizational, and communications environments. Production processes shaped by network technologies have created demand for 'knowledge workers' with higher educational pre-requisites and training needs than previously required in routine workers. The 'fast company' environment of the twenty-first century imposes accelerated product cycles and rapid turnover in knowledge and skills, giving rise to the idea of the 'learning organization' served by its own university (Lester, 1998). As June Delano, Director of Eastman Kodak's executive training unit states it, 'our ability to learn is our ability to compete' (Greco, 1997). Re-engineering has reduced firms' core employees, and increased partnerships and alliances with other firms networked in the value chain. Mergers and acquisitions also brought employees from two or more firms together. Corporate universities have been relied upon to update core skills, create a common vision of business context and a common culture across a value chain, or within a newly merged firm (Authers, 1999; Meister, 1998). Re-engineering has also generated an increase in short term employment, and labour unions have sought to increase the 'employability security' of technical workers by approaches that permit them continually to upgrade their skills and earn university diplomas (Lester, 1998).

Corporate universities are as diverse as their sponsoring firms. But as the phenomenon matures this diversity is decreasing. Distinct goals, methods, patterns of organization and curricula are becoming standard (Authers, 1999). Corporate universities are breaking off as a distinct academic culture, a shadow institution.

Corporate universities as institutional fragments. When corporate universities began operating in the 1980s and 1990s they frequently sought alliances with established universities. But they ran into conflicts with the academic culture that were similar to those encountered by John Sperling's adult programs. Departmental and collegial committees resisted the corporate focus and were in any case unable to respond with sufficient flexibility and speed to changing curriculum demands. Further, the education and training provided by professors continued to be general and academic, even when shaped through consultation with corporate university leaders. As a result, these early efforts proved unsuccessful, and many firms turned to new for-profit organizations to serve their education and training needs.

Corporate universities shape learning opportunities to support the strategic goals of their firms (Fenn, 1999; Meister, 1998). Many sponsoring firms are large multinational corporations with global reach. Unlike geographically specific universities, corporate universities have no 'public' mission tied to specific states, provinces, regions or nations. Rather, the organizational mission is tied closely to the firm and its 'value chain', its suppliers, partners and customers, often on a global scale. The primary curricular emphases have been global management, information network technologies, cross-functional problem solving, teamwork training and efforts to establish a common culture. Most are virtual or hollow organizations retaining only command and control functions. They out-source registry, curriculum design, instruction, library support, assessment, and degree conferral to organizations with which they have long-standing relations (Morrison, 2002; Moore, 2002). This allows the corporate universities to respond rapidly to changing demands, 'ramping up' or 'ramping down' without internal friction.

Few corporate universities grant their own diplomas or graduate degrees. A recent study found only five accredited degree-granting corporate units, with seven more working to achieve candidate status. Combined, these twelve organizations accounted for only 11,000 students (Allen, 2002, p. 5). The number remains low for two reasons. First, when corporate universities achieve accreditation, they often spin off from the parent corporation and operate as free-standing firms. Second, most corporate universities interested in degree-granting programs wish to avoid the hassles of licensure and accreditation. They form alliances with accredited universities, established non-profit universities or for-profits, including educational service corporations focused exclusively on the corporate university market.

One example is International Management Centres Association (IMCA) a firm serving the corporate universities of many large trans-national firms (Prestoungrange & Oliver, 2000). IMCA was born in 1964 when some innovative young British MBAs concurred that management education had become too academic, and set out to associate it more closely with management practice. They first sought affiliation with the British Institute of Management, which sponsored academic degrees through a national committee. They also proposed that the British Institute publish a practice-directed journal modelled on the Harvard Business Review. The Institute refused both requests. The young scholars were compelled to form their own Institute and Journal, now named Management Decision. Later they initiated new journals to publish work in emerging management fields such as technology and knowledge management. Today, they publish over 150 academic e-journals distributed through Emerald, a leading database of management scholarship. In 1982 IMCA began to offer its own practical diplomas and graduate degrees, based on action learning, the idea that the best 'school of management' is the manager's own workplace. Reginald Revans, a leading exponent of action learning and a respected business professor who had dropped out of academia in the 1960s, became president of IMCA.

IMCA provides tutoring services and accredited degree programs for corporate universities. Like the University of Phoenix, IMCA has a distinct teaching-learning method. Its programs dispense entirely with pre-set curricula. Students work in virtual 'sets', classes with classmates distributed globally, on individual projects within their assigned areas of work responsibility. Each student is answerable to his or her immediate work manager for both work assignments and assessment of firm-related dimensions learning. The members of a 'set' often hold parallel positions in organizational units of the firm located in different nations or regions, for example as managers of catering departments of hotels in the same chain, and thus have much to share with one another other. IMCA tutors, serving the distributed sets online, guide student reflection (including study of the best and most relevant academic literature) on work problems and decisions. Reflections generate both better management behaviours and academic deliverables, written studies that meet the knowledge needs of their classmates, their firms, and other professional managers, and also meet the standards of academic peer-review.

The most striking deviation from standard academic practice is the rejection of curriculum as a design for learning. Unlike the University of Phoenix, where courses depend on strict adherence to normative curricula, IMCA regards such curricula as no more than reference points which can easily become 'dangerous impositions' (Prestoungrange, 2000, p. 6). Students learn more from one another in informal exchanges than from formal instruction. Once a curriculum is dispensed with, students in the 'set' are free to compare problems and approaches with one another, while receiving guidance from academic tutors with credibility in the field of practice not simply the field of knowledge per se (p. 7). Their learning is assessed by authentic rather than formal criteria: that is, its contribution to the strategic mission of the firm and the professional literature of management.

Corporate universities as imperfect copies. Traditional universities and academic professionals have not accepted corporate universities as authorized higher education organizations. Instead they are 'feared and reviled,' regarded as intruders 'poaching on academic turf,' and rejected for their anti-academic practices such as treating the knowledge they generate as intellectual property of the firm (Allen, 2002, pp. 5 & 7). McDonald's Hamburger University, Toyota University's Department of the Toyota Way, and MGM Grand Hotel's University of Oz with its own ThD, its Doctor of Thinkology degree, have simply been ridiculed.

IMCA degree programs were actually outlawed. From the beginning they succeeded financially both in Britain and internationally, but met with hostility and sustained opposition from the business school establishment for IMCA's determination to offer its own degrees. In 1988 the Thatcher government passed legislation making all diploma programs subject to approval by the universities, 'the very institutions IMCA had been established to contradict' (Prestoungrange & Oliver, 2000, pp. 91-92). The IMCA degree programs were declared illegal. Refusing to subordinate itself to the mainstream academy, IMCA re-structured itself as a trans-national educational corporation, operating outside Britain in Australia, New Zealand, the United States, Singapore, Canada, Finland, the Netherlands, Curacao, and South Africa. It offers academic degrees through its University of Action Learning, now known as Revans University.

For their part IMCA leaders have regarded mainstream higher education as an illegitimate cartel organized to hold back external pressures for change by limiting supply and creating insurmountable barriers to entry (Prestoungrange, 2000, pp. 4-5). They see the Thatcher legislation of 1988 as a betrayal of free market principles and have adopted the mission of breaking the cartel and opening higher education to market competition (Prestoungrange, 2000, 5; Watson 2000, p. 26).

Corporate university divisions obscured by established universities. Corporate university divisions of non-educational firms are largely invisible, enclosed within parent firms with little media coverage. Degree granting educational suppliers like IMCA are no better known to the public or the academic profession than are, for example, suppliers of specialized capital equipment.[31]

Though established universities provide services for corporate universities, their role as corporate education suppliers remains largely unknown.

[31] In April 2003 I visited the IMCA headquarters, located prominently in Oxford, England across from the Ashmolean Museum. I then asked several members of the Department of Education at Oxford University if they knew about this organization and not one did.

Corporate universities as omens of the future. Nonetheless, once novel the corporate university is rapidly spreading. Motorola University is a good example. Among the American pioneers in the 1980s, it now offers Motorola's 90,000 global employees a choice of 10,000 courses, and is central to the firm's learning culture and knowledge management system. Small as well as large firms are creating their own corporate universities (Fenn, 1999). In the United States there are approximately 2000 of these units. In the 1990s, corporate universities gained a foothold in the UK, serving employees of Hilton, Lloyds TSB, BAE Systems, the BBC, Rolls-Royce, BT and IBM, among others (Cottee, 2003). Corporate universities spread in Europe (Renauld-Coulon, 2002) and Asia Pacific (Dickson, 2002), especially since 1998. They are most developed in 'post-industrial' industries such as finance, healthcare, telecommunications, entertainment, utilities, but are also making headway in traditional industries such as retailing (Sears University). The United States Navy (Ellison, 2001) and the British National Health Service (Sanders, 2002) recently opened in-house universities. The model works for public as well as private sector organizations as these move to re-tool themselves for a more global and high technology environment.

Corporate universities provide an entirely different product to that of established universities. They are 'positive subverters of present practices' (Watson, 2000). They are already impacting the higher education institution in recruitment and retention, academic standards, and pricing, and this influence will grow.

Like adult university corporations, corporate universities compete with mainstream organizations for non-traditional students, and increasingly will compete for traditional students as well. They offer the promise of career-relevant learning without deferring employment or incurring tuition costs and debts. As the London *Independent* put it: 'In today's topsy-turvy world of downsizing and short-term contracts, the long-term prospects that companies with universities potentially provide may be far more of a lure for a young person than a dry and dusty mainstream university course' (Nurden, 1999). The Committee of Vice Chancellors and Principals, representing more than 100 British universities, has warned that competition from corporate universities represents 'a key challenge facing the UK university system (Targett, 1999).

The educational standards of accrediting agencies and professional bodies are shifting to respond to the changing educational needs of firms, as addressed by corporate universities. The American Assembly of Collegiate Schools of Business (AACSB), the business school body, recently updated its accreditation criteria. According to Bill Laidlaw, Executive Director, 'basically we don't care how many books you've got in the library and the qualifications. What we care about is your mission, and how you are managing your resources to fulfil those objectives' (Authers, 1998). The Arthur D. Little School of Management in Cambridge, MA., was the first non-traditional school to be accepted into pre-candidacy for accreditation by the AACSB (Greco, 1997).

Mainstream universities partnering with corporate universities gain access to the firm's large base of employees as potential long-term customers of educational services. But as purchasers of large blocks of academic credits, firms can negotiate steep volume discounts (Mottl, 1999). Mainstream organizations have to price standard degree courses competitively to maintain market share for corporate educational services. The MBA, once a highly profitable product, is now seen as a 'loss leader' (Authers, 1998). But education services are increasingly sold as commodities, and price competition is stiff. For-profit educational corporations, without heavy brick and mortar overhead, can out bid mainstream universities and compel them (like established airlines challenged by low cost carriers) to form their own low cost, low overhead divisions that copy for-profit suppliers like IMCA. The DeakinPrime division of Australia's Deakin University is a prominent example (Dickson, 2002). Corporate universities are emerging from the shadows.

> If institutions of higher education wish to remain viable forces in the private sector ... they will no doubt be forced to rethink their relationship to business and overhaul their methodologies, products, services and delivery vehicles. Essentially this means they will have to behave more like businesses and treat their . . . adult students and their sponsoring companies, as valued customers (Kottke, 1999).

REFERENCES

Allen, M. (Ed.) (2002). *The corporate university handbook: Designing, managing and growing a successful program*. New York: Amacom.

Aronowitz, S. (2000). *The knowledge factory: Dismantling the corporate university and creating true higher learning*. Boston: Beacon Press.

Authers, J. (1998) A world of opportunities: Corporate universities are the newest currency but they may well pose a threat to smaller business schools. Financial Times (London), October 5.

Authers, J. (1999) Keeping company with the campus: Recent mergers have enhanced the role of corporate universities. *Financial Times* (London), April 26.

Barley, K. (2002). Corporate university structures that reflect organizational structures. In M. Allen, (Ed.), *The corporate university handbook: Designing, managing and growing a successful program*. New York: Amacom.

Clark, B.R. (1998). *Creating entrepreneurial universities : Organizational pathways of transformation*. Oxford: Pergamon.

Cottee, P. (2003). University challenged, *The Guardian*, October 20, 4.

Currie, J; DeAngeles, R., de Boer, H., Huisman, J. & Lacotte, C. (2003). *Globalizing practices and university responses: European and American differences*. Westport, CN. Praeger.

Dickson, I. (2002). Corporate universities in Australia and Southeast Asia. In M. Allen, (Ed.), *The corporate university handbook: Designing, managing and growing a successful program*. New York: Amacom.

Eisenstadt, S. (1972) Social institutions: the concept, in D. Sills, (Ed.) *The encyclopedia of the social sciences, Volume 7*, 409-421. New York: Macmillan.

Ellison, D. (2001). The Navy has its own corporate university. *Proceedings of the United States Naval Institute*, 127 (9), 74-5.

Fenn, D. (1999). Corporate universities for small companies. *Inc.*, February.

Greco, J. (1997) Corporate home schooling. *The Journal of Business Strategy*; 18 (3), 48-52.

Hamilton, W. (1932). Social institutions, in E. Seligman (Ed.) *The encyclopedia of social sciences. Volume 8*, 84-89. New York: MacMillan.

Hopkins, J. (2004). Sylvan revenue jumps 43%, operating income triples in 4Q; Firm that quit tutoring 'thriving beyond belief". *Baltimore Sun*, February 20, 1D.

Kottke, J. (1999). Review of J. Meister's Corporate universities: lessons in building a world-class work force. *Personnel Psychology*, 52 (2), 530-533.

Krapels, R.; Ryan, C. & Lane, J. (1998). Education initiatives inside business today. *Business Communication Quarterly*, 61, December, 4.

Lester, T. (1998). The in-house campus. *The Times* (London), November 26.

Meister, J. C. (1998). *Corporate universities: lessons in building a world-class work force* (revised). New York: McGraw-Hill.

Moore, J. (2002). Running a corporate university like a business. In M. Allen, *The corporate university handbook: Designing, managing and growing a successful program*. New York: Amacom.

Morrison, M. (2002). Creating a corporate university. In M. Allen, *The corporate university handbook: Designing, managing and growing a successful program*. New York: Amacom.

Mottl, J. N. (1999). Corporate universities grow. *Internetweek*, 15 March, 756.

Nurden, R. (1999). Corporate universities threaten the old guard. *The Independent* (London) 18 April, 3.

Prestoungrange, G. (2000). Customer orientation and motivation. In Prestoungrange, G., Sandelands, E. & Teare, R. (Eds.) *The virtual learning organization: Learning at the workplace campus*. London: Continuum.

Prestoungrange, G. (2003). *Back and forth*. East Lothian, Scotland: Prestoungrange University Press.

Prestoungrange, G., Sandelands, E. & Teare, R. (Eds.) (2000). *The virtual learning organization: Learning at the workplace campus*. London: Continuum.

Prestoungrange, G. and Oliver, (2000). International Management Centres as a learning organization.

Prestoungrange, G., Sandelands, E. & Teare, R. (Eds.) (2000). *The virtual learning organization: Learning at the workplace campus*. London: Continuum.

Readings, B. (1996). *The university in ruins*. Cambridge, MA: Harvard University Press.

Ruch, R. (2001). *Higher Ed, Inc.: the rise of the for-profit university*. Baltimore: The Johns Hopkins University Press.

Sanders, C. (2002). NHSU chief outlines future vision. *Times Higher Education Supplement*, 1523, 3.

Slaughter, S. and Leslie, L. L. (1997). *Academic capitalism: Politics, policies, and the entrepreneurial university*. Baltimore : Johns Hopkins University Press.

Sperling, J. (2000). *Rebel with a cause: The entrepreneur who created the University of Phoenix and the for-profit revolution in higher education*. New York: John Wiley & Sons.

Targett, S. (1999). Degree of discomfort in the corporate university challenge, Financial Times (London) 20 January, 10.

Turner, J. (1997). *The institutional order*. New York: Longman.

Leonard J. Waks
Temple University
Philadelphia, United States of America

JANE KENWAY, ELIZABETH BULLEN & SIMON ROBB

5. THE KNOWLEDGE ECONOMY, THE TECHNO-PRENEUR AND THE PROBLEMATIC FUTURE OF THE UNIVERSITY

INTRODUCTION

Knowledge economy polices and their associated innovation systems exert a major influence over university research around the world and will have far reaching consequences for universities of the future. These policies and systems position universities and knowledge, the university's stock in trade, as central to economic growth. They also make provision for more research funding. This combination of flattery and finance is seductive for universities. Few have questioned the underlying logic of the policy package and the long-term implications for universities and their academic staff, let alone for research and knowledge.

Our purpose in this chapter is to lay out the key features of this logic and to take it to its logical extremity. In so doing we highlight the sorts of academics this package seeks to produce and the possible problems that this may engender. Our overall argument is that current knowledge economy polices and innovation systems tend to ignore the distinctive features of universities and scholarly communities and that, in so doing, they put in peril aspects of what they seek to achieve, and much else besides. We begin by describing one trajectory in economic theory in order to focus on national innovation systems as a means of explaining the significance of universities in the knowledge economy. We then identify the key features of innovation systems and their implied ideal human subjects in the academy. Finally, we offer a critique of both.

ECONOMICS AND INNOVATION

It was Schumpeter who first theorized innovation as a key factor of economic growth. 'The capitalist engine', he claimed, is kept 'in motion' by the 'new consumers' goods, the new methods of production or transportation, the new markets, the new forms of industrial organization that capitalist enterprise creates' (Schumpeter, 1943, p. 83). These are products of the innovation process, which Schumpeter defined as a 'combinatory' activity involving new combinations of existing resources. Innovation differs from invention, new ideas or novelty, however, in that it leads to commercialization. Innovation, comments Fagerberg (2002, p. 9), is 'a specific social activity (function) carried out in the economic

Simon Marginson (ed.), Prospects of Higher Education, 121–137.

sphere with a commercial purpose'. According to one commentary Schumpeter regarded innovation as a driver of economic growth because it

> created new opportunities for profits, which in turn attracted a 'swarm' of imitators and improvers to exploit the new opening with a wave of new investment, generating boom conditions. The competitive processes set in motion by this 'swarming' then gradually eroded the margins of innovative profits (as in Marx's model), but before the system could settle into an equilibrium condition the whole process would start again through the destabilizing effects of a new wave of innovation (Freeman et al., 1982, p. 19).

This route is famously described as 'creative destruction' which, linked with technical innovation and long wave business cycles, refers to the 'process of industrial mutation ... that incessantly revolutionizes the economic structure *from within*, incessantly destroying the old one, incessantly creating a new one' (Schumpeter, 1943, p. 83, italics in original).

This particular dimension of Schumpeter's thought has been further developed by the neo-Schumpeterian school of post-Fordism originating in the Science Policy Research Unit (SPRU), University of Sussex. There Christopher Freeman and his colleagues drew on Schumpeterian evolutionary economics, in particular the notion of innovation as a driver of economic growth, to inform their attempts to theorize the dynamics of technology, growth and trade in the 1980s (see for instance, Freeman, 1987, Freeman et al., 1982, Freeman & Perez, 1988). Their work has influenced the conceptualization of the knowledge economy in two key respects.

First, it associates the characteristics of the social and economic change that occurs during each business cycle or Kondratiev with the dominant 'factor' industry or technical innovation which characterizes it. For instance, the fourth Kondratiev or long wave was Fordist. It was typified by mass production and consumption, and its dominant factor industry was electro-mechanical technology. In the current fifth Kondratiev, the major factor industry is microelectronics and the 'key 'carrier' sectors' include computers and software, telecoms, computer integrated manufacturing/new materials, information technology (IT) services, biotechnology, space/satellite and environmental technologies (Perez, 1985). It is this emphasis on technological innovation as a driver of economic growth which underpins the emphasis on techno-scientific knowledge in the knowledge economy. The second key contribution to the knowledge economy is the conceptualization of national innovation systems – 'networks of institutions in the public and private sectors whose activities and interactions initiate, import, modify and diffuse new technologies' (Freeman, 1987, p. 4). As the OECD (1996, p. 7) explains, 'The configuration of national innovation systems, which consist of the flows and relationships among industry, government and academia in the development of science and technology, is an important economic determinant'.

More recently, new growth theory (Romer, 1990; 1994) has reinforced these views. Romer also contends that economic growth is driven by technological progress or innovation. He argues that it involves the input of existing knowledge and human capital to make new and improved knowledge products. Also known as endogenous growth theory, new growth theory differs from classical economic theory, which acknowledges the importance of knowledge to economic growth but regards knowledge as exogenous – that is, external to – the economic process or growth model (Solow, 1970). In endogenous models of macro-economic growth, knowledge is internal to the model, and growth is the result of maximizing the 'behaviour' of knowledge workers and knowledge resources. National systems of innovation operate to facilitate knowledge flows that will accelerate technological innovation and orient knowledge production to commercial application. They operate to coordinate knowledge production with macro-economic goals. They are also designed to maximize the behaviour of university knowledge workers.

National innovation systems policies are reshaping knowledge production and exchange in and beyond the university in many countries around the world including Australia, the United Kingdom and other EU countries, Singapore, Korea and India. According to a recent OECD (2003, p. 107) report, higher education reform in New Zealand, for example, 'sets out a five-year approach for a more collaborative and co-operative tertiary system, contributing to national goals and more closely connected to enterprises and local communities'. The Danish Ministry of Science, Technology and Innovation was established in 2001 'to enhance interaction between business and the worlds of research and education' and 'A new university bill introduced a reform of governance designed to enhance universities' exchange of knowledge with economy and society' (OECD, 2003, p. 104). A 2003 Australian higher education reform document indicates the new positioning of the university within this framework when it states, 'Research and innovation play a vital role in building Australia's competitive strength in a global knowledge-based economy. Universities clearly have a central role to play as major producers of basic and applied research' (Nelson, 2003, p. 31). Universities have become central to innovation systems and so, too, potentially, have their academic staff assuming they can be reshaped in the appropriate manner. The knowledge economy generally and innovation systems particularly require particular sorts of human subjects and it is to these we now turn.

INNOVATION SYSTEMS AND THEIR IDEAL ACADEMICS

The innovation systems associated with the knowledge economy policies typically promote particular versions of techno-science, knowledge networks and research commercialization. Let us consider each in turn and the types of academic they seek to produce. We should explain here firstly that although we identify three types of ideal academic subject, they overlap and intersect to become the techno-preneur. The second point is that these subjects are preferred by and implied in policy texts and in the economic imperatives that inform them. Elsewhere we offer detailed accounts of the various ways in which academics and other university staff adopt or resist these ideal subject positions (Kenway, at al., 2006).[32]

Techno Scientism and the Techno-scientist

The flow-on effects of knowledge economy policies and innovation systems for the university are indicated by such things as the provision of incentives to increase enrolment in the so-called 'enabling' sciences (mathematics, physics and chemistry) to feed the applied sciences; the promotion of ICT skills and online learning; and the techno-economic orientation of national research priorities. National research priorities typically promote university research in those areas seen to characterize the fifth Kondratiev - information technology (IT) services, biotechnology, space/satellite and environmental technologies (Perez, 1985).

We provide an Australian example, but note that innovation policy developments in Australia are similar to those in many other parts of the world (see for instance, the Canadian innovation strategy, *Knowledge matters: Skills and learning for Canadians* (Government of Canada, 2002) and the UK white paper, *Our competitive future* (Department of Trade and Industry UK, 1998).

In 2001 the Australian federal government released *Backing Australia's ability: An innovation action plan for the future*. This was the outcome of a series of innovation policies and its aim was to provide effective support for the 'three key elements in the innovation process', which begins with 'strengthening our ability to generate ideas and undertake research' (Commonwealth of Australia, 2001, p. 14). Among an array of economic, taxation and education reforms, *Backing Australia's ability* makes provision for greater research support via grants, salaries and infrastructure; the creation of centres of excellence in information and communication technology (ICT) and biotechnology; funding and tax incentives designed to encourage R&D in industry and small enterprise; and increased education spending with a priority on science, mathematics and technology. *Backing Australia's ability* also flagged the development of national research priorities. When first announced, these were almost totally focused on science and technology.[33] It was only through persistent lobbying by the Australian academies

---

[32] What we find when we interview those involved in this processes is that they interpret knowledge economy/innovation imperative in various ways which are assembled from bits of economic necessity, government policy as well as practiced diversity and sectional or disciplinary self-interest.

[33] Announced in December 2002 these were
An environmentally sustainable Australia (priority goals include water, salinity, carbon emissions, biodiversity, deep earth resources);
Promoting and maintaining good health (priority goals are early childhood health, ageing and preventative healthcare);
Frontier technologies for building and transforming Australian industries (priority goals are breakthrough science (bio-informatics, quantum computing, geo-informatics, nano-assembly), frontier technologies (biotechnology, nanotechnology, ICT, photonics, genomics/phenomics and complex systems), advanced materials (biomaterials, ceramics, polymers, light metals, smart materials) and smart information use (e-data management and creative use of digital technologies).

of the humanities and of the social sciences that the National Research Priorities were amended to include one priority goal for each of the four chief priorities that offered scope for contributions from the non-science disciplines.[34] We note, however, that the function of these socially-oriented priority goals is arguably to deal with the impacts of, and remove the barriers to, the government's techno-scientific goals. 'Promoting an innovative culture and economy' is a case in point.[35]

This strong techno-scientific orientation was further reinforced by related policy and funding initiatives in Australia. In addition to the AUD $1.5 billion Department of Education, Science and Training annual budget allocated via the Australian Research Council (ARC) and formula-driven research funding schemes, *Backing Australia's ability* allocated an additional $3 billion funding package for national science and innovation over five years to university research and research training for the period 2002 and 2006 (Nelson, 2003, p. 119). In 2003, 34 per cent of the available ARC funding was allocated to its priority funding areas: nano-materials and bio-materials, genome/phenome research, complex/intelligent systems and photon science and technology (Nelson, 2003, p. 96).

In constructing university research as mainly techno-scientific research, knowledge economy policies construct university researchers as techno-scientists. The techno-scientist is not the same as the scientist, who may engage in pure or applied research. The techno-scientist presumes a much narrower subjectivity that combines scientific rationally with an instrumental and opportunistic sensibility. We can infer from OECD (1996, pp. 21-25) that the techno-scientist sets out to solve technological problems identified by industry and in the process surrenders any pretence to intellectual autonomy and any responsibility to undertake basic or curiosity-based research. He or she would not, for instance, consider science a 'public good', but rather as a tradable commodity. This privileging of the techno-scientific subject encourages academics in all disciplines, even in those disciplines marginalized by innovation systems, to restyle themselves in the manner prescribed or to risk being seen as academically and economically redundant.

Commercialisation and the Entrepreneur

According to a certain line of thinking in the economics of knowledge (Hayek, 1937, Kirzner, 1973), innovation of necessity involves entrepreneurial activity. Kirzner (1984, online) sums up the entrepreneur as 'the agent that spurs society to take advantage of existing scattered and dispersed knowledge' and 'generates and harnesses new technological knowledge, and discovers entirely new bodies of resources that had been hitherto overlooked'. Schumpeter also regards the

---

Safeguarding Australia (priority goals relate to critical infrastructure, new defense technologies, surveillance systems and application of new technologies to counteract invasive species

[34] The enhanced priorities were announced by the Minister for Education, Science and Training on 28 November 2003, 11 months after the Prime Minister announced the first set.

[35] For a discussion of the relationship of the knowledge economy to the arts and humanities see Kenway et al. (2004a).

individual entrepreneur as the key agent of innovation, but was prescient in predicting that innovation would become endogenized as a result of both the incorporation of research and development within firms and of collaboration between research institutions and industry, creating a "bureaucratized" type of innovation' (Freeman et al., 1982, p. 41). As this suggests, innovation becomes an orchestrated process in which 'The "coupling" between science, technology, innovative investment and the market, once loose and subject to long time delays, is now much more intimate and continuous' (p. 41). Further, according to the OECD (1996, p. 22), increased public funding of research will 'increase the variety of knowledge that might eventually find its way into commercial application'. The entrepreneur mediates between public funding and commercial application.

There is an entrepreneurial emphasis in knowledge economy policies directed at both 'developed' and 'developing' nations. In so-called developed nations this is evident in the policy and research initiatives of the OECD and the European Union (EU). Both give the entrepreneurial subject high priority. The EU has, for example set itself the goal of becoming the 'most competitive and dynamic knowledge-driven economy in the world' a goal, which it asserts, can only be achieved 'by making Europe more entrepreneurial and innovative' (European Commission, 2000, p. 2). The shorthand in operation here is typical; the production of the entrepreneurial subject is equated with future prosperity. Within this scenario,

> General knowledge about business and entrepreneurship needs to be taught, right through primary, secondary and tertiary education. Enterprise policy will aim at making specific enterprise and business-related modules or programmes an essential ingredient of education schemes at secondary level and at colleges and universities. For a scientist or an engineer, being able to draft a business plan should be as natural as doing a scientific experiment or writing a theoretical article (European Commission, 2000, p. 3).

In a recent paper presented to the OECD on the entrepreneurial university some of its key tasks are articulated. The entrepreneurial university:

- responds to varying student needs and circumstances
- takes account of labour market requirements and employer needs
- embeds entrepreneurial skills and ethical values in course offerings
- develops application linkages for research
- undertakes collaborative research with industry
- participates in research commercialisation ventures
- establishes diverse sources of income
- provides commercially valuable services
- plans for growth in total income
- competes successfully in its markets
- collaborates with others for full service delivery
- employs flexible staffing strategies
- manages intellectual property strategically (Gallagher, 2000, p. 2).

It is not difficult to see those institutional tasks as key facets of an entrepreneurial subject within the university. Evident here is an elaboration of the techno-scientific subject, with a particular emphasis on the commercialization and managerialization of research and teaching.

Two recent policy statements by the World Bank indicate how the entrepreneur is constructed for the so-called developing nations. The higher education entrepreneur is the ideal subject who, by commercializing new knowledge not only contributes to general economic growth, but who also increases the flow of capital into the university itself.[36] World Bank policy (World Bank, 2002a, 2002b, 2003) with respect to higher education is typically concerned with the governments of developing countries withdrawing from the financing of tertiary education, except where there is a need to subsidize disadvantaged students. Governments therefore need to stimulate private investment in higher education and to collaborate with private investors and to ensure their profit. Further, higher education is unambiguously expected to deliver human capital for the economy. The characteristics of the human capital required are typically adaptability, creativity, flexibility and the ability to innovate. World Bank policy indicates that these are essentially products of a de-regulated education market. Education is only valuable when it is tied to an overall innovation plan, accommodates the interests of industry both at the level of content and mode of delivery, and when it is funded in a way that maximizes competition and hence the opportunities for investment and private gain. *Building knowledge economies* argues that

> Continuous, market-driven innovation is the key to competitiveness, and thus to economic growth, in the knowledge economy. This requires not only a strong science and technology base, but, just as importantly, the capacity to link fundamental and applied research; to convert *the results of that research to new products, services, processes, or materials; and to bring these innovations quickly to market*. It also entails an ability to tap into and participate in regional and global networks of research and innovation. (World Bank, 2002b, p. 21, our emphasis).

The World Bank is fostering the growth of a higher education entrepreneurial personality within developing and 'transitioning' nations. Demands on developing nations to restrict public funding to primary and secondary education put particular pressures on the higher education sector to produce the entrepreneurial subject.[37]

Knowledge Networks and the Knowledge Networker

National systems of innovation are a key mechanism for facilitating endogenous growth in the knowledge economy. Innovation becomes an organizational process

---

[36] For a discussion of the knowledge economy, the entrepreneur and public education in Australian schools see Robb et al. (2003).

[37] For an elaboration of the knowledge economy and developed and developing nations see Kenway et al. (2003b).

based on knowledge networks whereby 'interactive learning involving producers and users in experimentation and exchange of information is the driver of innovation' (OECD 1996, p. 14). Innovation systems seek to bring together top researchers and to facilitate knowledge flows between them to produce the leading edge research. They also seek to bring together different 'players' in the innovation system - government, academe and industry. If, as Schumpeter says, innovation entails new combinations of existing resources, then this interaction maximizes the number of possible combinations and varieties of knowledge.

Research links with industry act to reinforce the commercial orientation of research noted above and to alter the conventional sequence of innovation which typically begins with new research and passes through the various stages of development, production, marketing, and ultimately the consumption of new products. In the knowledge economy this process is no longer linear. As the OECD (1996, p. 14) explains, 'innovation can stem from many sources' and thus 'requires considerable communication among different actors – firms, laboratories, academic institutions and consumers – as well as feedback between science, engineering, product development, manufacturing and marketing'.

Such thinking is reflected in research policies that promote the creation of research 'clusters' and 'centres of excellence' to assist with the generation of new knowledge and critical mass, and the formation of disciplinary, transdisciplinary and transnational networks to assist with the production of and access to the 'best' knowledge. For example, the Sixth Framework Program (2002-2006) of the European Commission which, incidentally seeks to drive 'European research further and faster than ever before' (European Commission, 2002, back cover), now funds what it calls 'Networks of Excellence'. It defines them thus:

> Networks of Excellence are designed to strengthen scientific and technological excellence on a particular research topic by integrating at European level the critical mass of resources and expertise needed to provide European leadership and to be a world force in that topic. This expertise will be networked around a joint program of activities aimed principally at creating a progressive and durable integration of the research capacities of the network partners while, of course at the same time advancing knowledge on the topic. (FP6 Instruments Task Force/European Commission, 2003, p. 1, emphasis in original).

Networks of Excellence are expected to engage with other research teams and with 'actors beyond the research community and with the public as a whole' in order to 'transfer knowledge' and 'spread excellence' and encourage 'take up activities' (European Commission, 2002, pp. 2-3).

This example is clearly about formally funded research networks, and it is worth noting that a number of national research councils now fund such networks (the Australian Research Council's 'research network' program and the 'priority networks' program of the Economic and Social Research Council, UK, are two such examples). However, the networking imperative goes beyond the formal imprimatur of funding bodies and is now systemic. In this context, relationships between academic staff, within and between universities, and between universities and the outside world are cast largely in commercial-contractual terms (Boden, Barker, Cox & Nedeva, 2004). There is considerable pressure on all academics to become particular sorts of networkers. The type of networker that innovation systems seek to produce sees all relationships in this commercial-contractual way. The policy documents noted above suggest that networkers are primarily interested in 'interactive learning', exchanging information across disciplinary or institutional borders spreading excellence and the frission that produces new ideas. Implicit in such policies is the notion that 'know who' is as important as 'know how' and 'know what' (OECD, 1996, p. 12). Not only is knowledge a tradable asset, so too are connections and relationships. These are conceived of instrumentally and commercially; in other words, in terms of use and exchange value. Other people or groups are evaluated in terms of what they can trade. They may be valued because they can back-fill the knowledge or the status that a particular knowledge network lacks and needs in order to competitively bid for resources. They may be valued because they have such things as insider knowledge about and direct links to key political and institutional figures and access to other valuable nodes and networks with various resources to disperse. Leverage and synergies are key words here.

THE TECHNO-PRENEUR, THE 'NOW' UNIVERSITY AND THE GIFT ECONOMY

It is 'inevitable that changing the relationships that frame knowledge production will change the nature of the knowledge produced' (Kenway, Boden & Epstein, 2004). And, of course, it is precisely the intention of knowledge economy and innovation policies to reshape the terms of universities and academics' future engagement with the economy and also society and culture. However, it is our view that such policies contain a number of un- or under-acknowledged problems which we will now discuss. Interestingly, some of these were alluded to and then implicitly dismissed by the OECD in *The knowledge-based economy* (1996).

A convenient condensation of the implied ideal subjects we discussed in the previous section can be found in the term techno-preneur, a subject who has networking skills, a techno-scientific orientation and an entrepreneurial sensibility. The techno-preneur can thrive in a competitive, increasingly privatized knowledge market, understands national economic needs, and meets these through profitable partnerships with national and international industry. The techno-preneur is concerned with on-going material gain through competitive means. It is our view that much about the techno-preneur mitigates against the intellectual work of universities and our task now is to explain how this is so.

As discussed elsewhere (Bullen et al., 2004b), the neo-Schumpeterian economics informing national innovation systems is criticized as technologically deterministic (Amin, 1994; May, 2002; Webster, 1995). In privileging the techno-sciences and the techno-scientific subject dictated by the knowledge economy, universities are also implicated in this technologically deterministic logic. In other words, they accord technology undue power as an agent of change, of the particular knowledges that are produced and taught in universities and the knowledges that languish as a result. Further, innovation policy has also been criticized because it tends to assume a 'developmental trajectory' and therefore tend to stress 'the independence of technology from social forces' and focus instead on the 'logical development of one innovation to the next' (May, 2002, p. 26). As Bimber (1995, p. 84) further explains, there is an assumption that 'technological developments occur according to some naturally given logic, which is not culturally or socially determined, and that these developments force social adaptation and changes'. They are positioned as forces over which people have no control, inviting passive accommodation to technologically-induced change (May, 2002). 'As a critical stage of advancement is reached, it becomes very difficult for national economies and individual firms to opt out of the new technological regime; they become "locked-in" to a universal developmental trajectory' (Elam, 1994, p. 45).

Much higher education reform designed to support knowledge economy imperatives, likewise locks the techno-scientific subject into this trajectory. From this perspective, it is not the techno-scientist who drives research and innovation, but the momentum of technology itself. Such a mind-set mitigates against a critical engagement with the logic of the policy itself. Hence the future consequences or risks of innovation as discussed by (Beck, 1992) and others are unlikely to be addressed. It is Beck's view that the hazards of techno-scientific progress are 'systemically grounded in the institutional and methodological approach of the sciences to risk' (Beck, 1992, p. 59). The sciences are therefore not only constrained in their ability to respond to risk, but implicated in the creation of risk. Clearly, risk is further heightened when the production of new knowledge is oriented to achieving immediately measurable impact and competitive advantage. According to the selection criteria used to set the Australian national research priorities, measurable impact refers to the ability to 'capture the benefits of research' or to further enhance the nation's 'innovation capacity by broadening the knowledge base and fostering acquisition of skills and "hot" research' (Department of Science Education and Training, 2003). The latter arguably makes innovation and economic growth ends in themselves.

Such imperatives do not merely fetishise the new as Brown and Michael (2003) argue, they fetishise 'the soon to be and a corresponding adjustment to the exploitation of emerging or future opportunities rather than established routines and habits ... Success depends on shortening time frames and, if possible, projecting them into the future'. The competitive environment associated with innovation policy then, has the effect of denying tradition, compressing time, and producing a sense of urgency with some already recognized unfortunate side effects. A recent article in *Nature* republished in an Australian newspaper (Pearson, 2004) reports on some of the ways the techno-preneur is evolving, noting examples of intellectual theft, cutting corners, and a reluctance to engage in the informal and free exchange of ideas. Taken to its extreme, research driven by innovative systems may discourage honesty, reflexivity and intellectual selflessness. Further, the goal of innovation is not the broader benefits it might bring, but the scientific 'breakthrough' itself, thus discouraging the sort of reflexivity needed to assess the risks of new technologies. Although bodies like the EU might encourage an ethical sensibility and a concern about the socio-economic impact of research, the system itself at best marginalizes such matters and at worst implicitly discourages them.

Even though knowledge economy policies and innovation systems dominate national and international policy higher education and research agendas, it should be noted that the techno-preneur is not the unproblematic darling of all intentional or supranational bodies. For instance, in December 2003 UNESCO hosted a forum on the knowledge economy which had at its starting point the narrow and reductionist logic of knowledge economy policies and hence of the promulgation of the techno-preneurial subject in universities around the world:

> The hegemony of neo-liberal ideology, grounded in the logic of the market, with privatization of the sphere of knowledge production as its advanced expression, has injected a perspective whereby current issues tend to be discussed largely in terms of managerial values and practices. In this setting, issues reduce to the economic aspect alone. They focus on the 'end application', on manpower training for employability and on wealth creation, spurred on by criteria of efficiency and by a market-driven rationale. Discussion couched in broader terms of scientific ends and purpose, of long-term development that can be sustained and of society's broader progress, figures little. (UNESCO, 2003)

While the techno-scientist is to be driven by the hidden hand of technology and by the push for applied knowledge, the entrepreneur is driven by all too obvious hand of the market. For the techno-preneur, the market and technology come together as the drivers of all else. The problem here is not so much that the knowledge generated and circulated in innovation systems will be in the context of application (Gibbons et al., 1994), but that the context of application in the knowledge economy almost always reduces to a commercial one. Commerce, education and research are not necessarily or always suitable bedfellows. The characteristics of the entrepreneurial university noted above offer a particularly shallow, short-term and sterile notion of teaching and research in universities Indeed, they exemplify a problem that the OECD warned against in 1996 about 'a potential conflict between knowledge production in the knowledge based economy and knowledge transmission or the primary, educational mission of the university' (OECD, 1996, p. 24). Certainly, innovation systems' obsession with intellectual property rights (IPR), product market regulations, competition rules, commercial in confidence and so forth exemplifies exactly this potential.

The 'tight coupling' between the university and business and industry, which the techno-prenueur is responsible for promoting, has other risks. These are not just for the circulation of knowledge, but also for its production and for the future of the university itself. As the OECD also acknowledged in 1996 (p. 25), if industry voice and investment have too much steering capacity, the 'contribution of academe to knowledge production may actually weaken under the burden of proving its economic relevance'. Thus the OECD acknowledges that there may be a conflict between the fundamental research mission of the university and immediate returns, between industry-directed and curiosity-based research. What it does not acknowledge is whether or not instrumentality and commercialization encourage abstraction and experimentation, or whether research in applied fields means that the knowledge that matters is limited by possible applications. As we have argued elsewhere (Bullen, et. al., 2004a), giving priority to applied research along with entrepreneurial activities tends also to give precedence to short-term commercial pay-offs, to dictate research priorities, to privilege corporate values over academic values in decision making, and to evaluate research performance in ways that create particular status and survival problems for disciplines not compatible with the commercial and entrepreneurial orientation of the innovation system. Arguably universities have a public responsibility to have a broad, rich and deep knowledge base which is attentive to a broad range of social and cultural knowledge and also to tradition. To residualize certain knowledges is to undermine the social contract between the public and the university in favour of those sectional interests with the money to pay for the knowledge they want. For these reasons and more, the techno-preneur is a highly problematic figure in contemporary university life.

Elsewhere, we have developed at length the argument that a gift economy is a repressed aspect of a knowledge economy and that a vibrant and generative intellectual community is, in many senses, dependent on a gift economy (Kenway et al., 2006). There is not the space to develop this argument fully here but a few key points are in order in the remainder of the paper. Commodity exchange alienates objects and subjects and ensures their flow in space and time free of tradition. Gift exchange binds objects and subjects in a symbolic and reciprocal relationship that de-limit their movements according to social and ethical codes. On the one hand, commodification facilitates freedom of the object and subject, but also leads to the destruction of the social and the ethical. The gift on the other hand, facilitates social bonds through non-commodified exchange. Gregory summarizes gift exchange as an 'exchange of inalienable things between transactors who are in a state of reciprocal dependence' (Gregory, 1982, p. 12). He suggests that this proposition is implicit in Marx and can be discovered by inverting the fundamentals of commodity exchange. He continues with this inversion as follows: 'commodities are *alienable* objects transacted by aliens; gifts are *inalienable* objects transacted by non-aliens' (1982, p. 43, emphasis in original). We might say that the techno-preneur desires to alienate all objects from a gift community and to turn the subjects of that community into aliens.

Marx asserted that commodity exchange moves from the margins to the 'interior of the community, exerting a disintegrating influence upon it' (Marx, 1971 [1859] p. 50, quoted in Gregory, 1982, p. 12). It can be argued likewise that commodity exchange exerts a disintegrating influence on the academic community and that the profiteering of ideas will have this fragmenting effect. No-one, for instance, will freely share with someone who is known to have an eye on a potential patent. Notions such as IPR and commercial-in-confidence are very different from notions of citation and acknowledgement and are at odds with the openness, the obligation to repay, and the sociality of the gift economy which has hitherto been an ideal feature of academic communities. Hyde suggests that intellectual communities emerge and are sustainable only when knowledge is circulated as a gift. As long as an intellectual community depends on the exchange of ideas (for example in journal publication), then it will also be dependent on gift exchange (1999, p. 83). Indeed, if commodity exchange is between free actors who engage in temporary relationships, as opposed to obligated actors who are engaged in an ongoing relationships (Carrier, 1995, p. 23), it follows that an intellectual community which depends on a continuity of ideas and relationships will be undermined by commodity exchange and by the actions of its ideal subject the techno-preneur.

So what of the free flow of ideas and the imperative to partner and collaborate associated with knowledge networks specifically and innovation systems generally? Does this not some how resemble a gift economy? It is not at all predictable how, or if at all, hyper-competitive performative knowledge networks will share with outsiders the knowledge and other benefits that accrue to them through their own network. Indeed, Intellectual Property regimes are among the most contentious aspects of these funded networks. The EU anticipates this sharing problem with a caveat against 'closed clubs' and with its insistence that each network 'be given a mission to *spread excellence* beyond the boundaries of its partnership' (FP6 Instruments Task Force Sixth Framework Program/European Commission, 2003, p. 1, emphasis in original). In this view, as noted earlier, others are conceived of as potential knowledge partners or beneficiaries. However, given the power of the commercialization meta-narrative, they are also likely to be seen as knowledge competitors, or as without use or exchange value. Either way they may become excluded from the flow of knowledge.

This is not likely to be a static scene however. Knowledge networks within innovation systems are based on pragmatics and, over time, will either be held together or break apart according to self-interest. And in this fast moving, short-termist environment, it may not pay knowledge techno-preneurs to invest too much time, money or psychic energy in particular networks because it is possible that they may be here today and gone tomorrow. Indeed, a sense of reciprocal bonds and tradition are an anathema to the techno-preneur. Innovation systems might try to normalize knowledge networks, but they also have the potential, paradoxically, to produce the individualistic footloose and promiscuous techno-prenueur, who in turn has the potential to undermine the ontological security of the academy which relies heavily on trust relationships and the bonds formed through reciprocity – in short, the gift of knowledge between academics.

The moral agendas associated with the networking imperative are ambiguous. They tap into but also contradict long-standing, even if somewhat at times mythical, gift traditions about the free and fair flow of knowledge within the academic community world-wide, not in trading blocks, the fruitfulness of such exchanges and about putting knowledge to work, or not, in various ways in and outside the academy. Conventionally such networks have been based on the notion of academic freedom, on a certain distance from that or those studied, and also a certain disinterest in the outcomes of that knowledge. These traditions have been seen as vital to the production of independent critical enquiry and to the ongoing production of new knowledge. In turn, this has involved an implicit social contract in which academics generate knowledge for various publics in exchange for public money and trust. The imperative to and possible consequences of 'tight coupling' with business and industry which we discuss above puts such trust at risk. It is not clear how university academics can guarantee their independence or disinterestedness in the face of such coupling practices.

In an intellectual community academics give to others the knowledge that has moved through them and has been transformed by them. They can give back to the community of knowledge what they have borrowed, or received for a time through the publication of research. Indeed, publication is a way for knowledge to circulate for the continuance of an intellectual community, in the same way that the gift circulates to maintain a community based on a gift economy. Hyde (1999 [1979]) discusses this notion in relation to scientific communities, acknowledging the work of Hagstrom (1975 [1965]). Hyde notes, for example, that papers for research publications are offered as gifts without payment yet they have a higher status than those in books which are commercially sold. We might also note that research papers circulate knowledge amongst the intellectual community and do so by working on and with the ideas of others. Research publications pass on ideas that have been re-worked, or refined, and are given back to the community. Research receives the gifts of others and then gives back in return. Likewise the debt that is owed to another is repaid through citation. Publications result in status, and prestige within the group, rather than in cash remuneration. Status and prestige are community defined positions which are dependent on the long-term viability of the community. Gregory discusses the notion of status in the following way:

> Commodity exchange relations are objective relations of equality established by the exchange of alienated objects between independent transactors. Gift exchange relations are personal relations of rank, established by the exchange of inalienable objects between transactors who are related (Gregory, 1982, p 71).

If a vibrant intellectual community is dependent on a gift economy then knowledge economy policy needs to be theorized in terms of the gift as well as the commodity. It needs to advocate on behalf of the one who gives, receives and repays knowledge for reasons of obligation to knowledge itself, solidarity towards an on-going intellectual community, and social and symbolic rewards such as rank and status. This applies to research and teaching. To paraphrase Gregory (1982, p. 19), an intellectual gift economy is a debt economy. The aim of a transactor in such an economy is to acquire as many intellectual-debtors as possible, not to maximize profit, as it is in the commodity economy. What a gift transactor desires is an intellectual community, the intellectual relationships that the exchange of gifts creates and not the things themselves. The ideal subject of knowledge economy discourse, the techno-preneur, finds such notions incomprehensible and even reprehensible. Debating if and how these two different subject positions might accommodate each other is vital to the future of the university.

REFERENCES

Amin, A. (1994). Models, fantasies and phantoms of transition, in A. Amin (Ed.) *Post-Fordism: A reader*. Oxford: Blackwell.

Beck, U. (1992). From industrial society to risk society: Questions of survival, structure and ecological enlightenment. *Theory, Culture & Society*, 9, 97-123.

Bimber, B. (1995). The three faces of technological determinism, in M.R. Smith & L. Marx (Eds.) *Does technology drive history: The dilemma of technological determinism*. Cambridge, MA: MIT Press.

Boden, R., Barker, K., Cox, D. & Nedeva, M. (2004). *Scrutinising science: The changing UK government of science*. Houndmills: Palgrave Macmillan.

Brown, N. & Michael, M. (2003). *A sociology of expectations: Retrospecting prospects and prospecting*. Technology Analysis and Strategic Management, 15 (1), 3-19.

Bullen, E., Kenway, J., & Robb, S. (2004a). Can the arts and humanities survive the knowledge economy? A beginner's guide to the issues. In J. Kenway, E. Bullen, & S. Robb *Innovation and tradition: The arts, humanities and the knowledge economy*. New York: Peter Lang.

Bullen, E., Kenway, J., & Robb, S. (2004b). 'Creative destruction'. Knowledge economy policy and the future of the arts and humanities in the academy. *Journal of Education Policy*, 19(1), 3-22.

Carrier, J. (1995). *Gifts and commodities: Exchange and Western capitalism since 1700*. London & New York: Routledge.

Commonwealth of Australia (2001). *Backing Australia's ability: An innovation action plan for the future*. Canberra: Commonwealth of Australia.

Department of Science Education and Training, Australia (2003). *Framework, national research priorities*. Accessed 15 March 2004 at: http://www.dest.gov.au/priorities/framework.htm

Department of Trade and Industry, UK (1998). *Our competitive future: Building the knowledge driven economy*. Accessed at: http://www.dti.gov.uk/comp/competitive/main.htm.

Elam, M. (1994) Puzzling out the post-Fordist debate: Technology, markets, institutions. In A. Amin (Ed.) *Post-Fordism: A reader*. Oxford: Blackwell.

European Commission (2000). *Challenges for enterprise policy in the knowledge-driven economy: Proposal for a Council decision on a multiannual programme for enterprise and entrepreneurship (2001-2005)*. COM(2000) 256 final/2. Brussels: Commission of the European Communities.

European Commission (2002). *Participating in European research: Sixth framework programme*. Luxembourg: Office for Official Publications of the European Communities.

Fagerberg, J. (2002). *A layman's guide to evolutionary economics*. Paper presented at conference on Industrial R&D and Innovation Policy. Leangkollen, Asker, 18-19 April.

FP6 Instruments Task Force/European Commission (2003). *Provisions for implementing networks of excellence*. European Commission. Accessed at http://europa.eu.int/comm/research/fp6/pdf/noe_120503final.pdf.

Freeman, C. (1987). *Technology policy and economic performance: Lessons from Japan*. New York: Pinter.

Freeman, C., Clark, J., & Soete, L. (1982). *Unemployment and technical innovation: A study of long waves and economic development*. London: Pinter.

Freeman, C. & Perez, C. (1988). Structural crises of adjustment, business cycles and investment behaviour. In G. Dosi, C. Freeman, R. Nelson, G. Silverberg, & L. Soete (Eds.) *Technical change and economic theory*. London: Pinter.

Gallagher, M. (2000). *The emergence of entrepreneurial public universities in Australia*. Commonwealth of Australia. Paper at IMHE General Conference of the OECD. Paris, September.

Gibbons, M., Limoges, C., Nowotny, H., Schwartzman, S., Scott, P., & Trow, M. (1994). *The new production of knowledge: The dynamics of science and research in contemporary societies*. London: Sage.

Government of Canada (2002). *Knowledge matters: Skills and learning for Canadians*. Hull: Human Resources Development Canada. Accessed at: http://www.innovationstrategy.gc.ca

Gregory, C.A. (1982). *Gifts and commodities*. London & New York: Academic.

Hagstrom, W. (1975 [1965]). *The scientific community*. Carbondale: Southern Illinois University Press.

Hayek, F. (1937). Economic of knowledge. *Economica* IV. Accessed at http://www.hayekcentre.org/friedrichhayek/hayek.html.

Hyde, L. (1999 [1979]). *The gift: Imagination and the erotic life of property*. New York: Random.

Kenway, J., Boden, R., & Epstein, D. (2004). Buiding networks. In: Boden, R., Kenway, J.,& Epstein, D., *The academics' support kit*. London: Sage.

Kenway, J., Bullen, E., Fahey, J., with Robb, S. (2006). *Haunting the knowledge economy*. London: Routledge.

Kirzner, I.M. (1973). *Competition and entrepreneurship*. Chicago: University of Chicago Press.

Kirzner, I.M. (1984). *The role of the entrepreneur in the economic system*. Paper presented at The Inaugural John Bonython Lecture, Adelaide, 30 July. Accessed at: http://www.cis.org.au/Events/JBL/jbl84.htm.

Marx, K. (1971 [1859]). *A contribution to the critique of political economy*. Moscow: Progress Publishers.

May, C. (2002). *The information society: A sceptical view*. Cambridge: Polity Press.

Nelson, B. (2003a). *Higher education report for 2003 to 2005 triennium*. Canberra: DEST/Commonwealth of Australia. Accessed at: http://www.dest.gov.au/highered/he_report/2003_2005/pdf/triennium2003_2005.pdf.

Nelson, B. (2003b). *Our universities: Backing Australia's future*. Canberra: Commonwealth of Australia. Accessed 17 February 2004 at: http://www.backingaustraliasfuture.gov.au/policy_paper/policy_paper.pdf.

Nelson, B. (2003c). *Higher education report for the 2004 to 2006 triennium*. Canberra: DEST/Commonwealth of Australia. Accessed at: http://www.dest.gov.au/highered/he_report/2004_2006/pdf/triennium_2004_2006.pdf.

Organization for Economic Co-operation and Development, OECD (1996). *The Knowlede-based economy*. Paris: OECD.

Organization for Economic Co-operation and Development, OECD (2003). *Education policy analysis*. Paris: OECD.

Pearson, H. (2004). Science suffers in race for 'scoops'. *The Weekend Australian*, 17-18 January.

Perez, C. (1985). Microelectronics, long waves and world structural system: New perspectives for developing countries. *World Development*, 13, 441-63.

Romer, P. (1990). Endogenous technological change. *Journal of Political Economy*, 98 (5), S71-S102.

Romer, P. (1994). The origins of endogenous growth. *The Journal of Economic Perspectives*, 8, 3-22.

Schumpeter, J.A. (1943). *Capitalism, socialism and democracy*. New York: Harper.

United Nations Educational, Scientific and Cultural Organization, UNESCO (2003). Knowledge society vs. knowledge economy. First global research seminar, Paris, 8-9 December. Accessed at: http://portal.unesco.org/education/ev.php?URL_ID=21012&URL_DO=DO_TOPIC&URL_SECTION=201.

Webster, F. (1995). *Theories of the information society*. New York: Routledge.

World Bank (2002a). *Constructing knowledge societies: New challenges for tertiary education*. Washington DC: Education Group Human Development Network, World Bank. Accessed at: http://www1.worldbank.org/education/pdf/Constructing%20Knowledge%20Societies.pdf.

World Bank (2002b). *Building knowledge economies: Opportunities and challenges for EU accession countries*. Final report of the knowledge economy forum 'Using knowledge for development in EU accession countries'. Paris: World Bank. Accessed at: http://lnweb18.worldbank.org/ECA/ECSSD.nsf/a3b026a6ee1e272585256ad2007130d3/9e9735587b22d64285256bce005ddbad/$FILE/Building%20Knowledge%20Economies-final%20final.pdf

World Bank (2003). *Lifelong learning in the global knowledge economy: Challenges for developing countries*. Washington DC: World Bank. Accessed at: http://www-wds.worldbank.org/servlet/WDS_IBank_Servlet?pcont=details&eid=000094946_03062104054940.

Jane Kenway, Elizabeth Bullen & Simon Robb
Faculty of Education
Monash University
Melbourne, Australia

PART II

PERSPECTIVES OF THE AMERICAS

'For there is room for only one centre at a time in a world-economy'

– Fernand Braudel, *The perspective of the world, Civilisation and capitalism*, Volume 3, FontanaPress, London, 1985, p. 33

PETER ECKEL, MADELEINE GREEN & KATE BERNIAZ

6. U.S. PROVIDERS AND PROGRAMS ABROAD:

A Proposed Cross-Border Framework[38]

INTRODUCTION

Metaphors about globalization are ubiquitous–'the world is getting smaller'; 'the world is flat' are common refrains in the popular press as well as academic circles. As one colleague quips, 'globalization' is quickly surpassing 'excellence' as the most overused term in the academy. Universities and colleges are very much affected by globalization, marked by its economic restructuring, network dependence, political reordering and heightened worker and student mobility and expectations (Van Damme, 2002).

Universities are both old hands and newcomers when it comes to globalization. They have long engaged in internationalization - 'integrating an international, intercultural, or global dimension into the purpose, functions or delivery of postsecondary education' (Knight, 2003a, p. 2) - and in cross-border education, the international mobility of students, faculty, researchers, programs and projects, and institutions. However, rapid developments of technology, competition for students and resources, changes in national policies, and the growth of new types of higher education providers, are changing the higher education landscape worldwide quickly and profoundly in untold ways. Van Damme (2002) predicts that globalization will generate new demands on universities to serve as knowledge centres, and increased student demand for higher and further education, creating a 'boom market' for higher education. He also envisions an erosion of national regulatory and policy frameworks as a result of borderless higher education.

The impact and implications of these issues on higher education are still uncertain, and will vary according to national context, particularly by the level of national economic development. For industrialized nations such as the United States, globalization poses relatively little threat. The developed nations are largely the beneficiaries of globalization in many spheres. In higher education they receive the vast majority of the world's mobile students, and more recently, export education through distance learning and off-shore programs. In developing nations, issues related to globalization are garnering significant attention by policy makers, scholars, and institutional leaders. For some, it is causing considerable alarm. The

[38] This paper is based on a previous paper (Eckel, P. D., Green, M. F., & Affolter-Caine, B. (2004) Curricular Joint Ventures: A New Chapter in U.S. Cross-Border Education? Policy Futures in Education. 2(2), 299-315) and ongoing work at the American Council on Education.

Simon Marginson (ed.), Prospects of Higher Education, 141–154.

values typically perpetuated by globalization, such as 'privatization' 'deregulation', and 'liberalization', are not necessarily compatible with national agendas in developing countries (Singh, 2002). The debate about the benefits or detriments of globalization are apparent in the controversies over the General Agreement on Trade in Services (GATS) negotiations of the World Trade Organization (WTO). Some believe that 'trade in education' is a positive globalizing force, providing access to higher education and meeting unmet demand, and fostering healthy competition among institutions. The sceptics point to the potential weakening of public education as a result of the proliferation of foreign educational investors and providers and their ability to recruit wealthy students into lucrative programs such as business and information technology (IT).

This chapter presents an emerging framework for understanding the variety of approaches being used by American colleges and universities to deliver degree programs outside the United States to non-U.S. citizens, an important and growing trend, although uncharted in any systemic way. This framework provides a structure for understanding how U.S. colleges and universities are leveraging their curricula internationally. We limit this framework to those international academic activities that result in academic degrees and exclude other more familiar joint international activities such as research collaboration and development work.

It is helpful to place these newer types of activities, those that provide U.S. degrees to non-U.S. students studying outside the country, in the larger picture of cross-border education. Knight (2003b) offers a useful typology for cross-border education that serves as an alternative framework for the more trade-oriented language of the four modes of supply used by GATS. Her framework suggests that cross border education can involve (1) *people* (students, faculty, or researchers going abroad for educational purposes); (2) *program mobility* (an educational program going abroad, such as distance learning or a joint academic program); (3) *providers* (an institution or other provider going or investing abroad for educational purposes); or (4) *projects and services* (partnership programs or commercial contracts for research, curriculum design, capacity building, etc.) The focus of the framework discussed in this chapter is on program and provider mobility.

U.S. HIGHER EDUCATION INTO THE GLOBAL ARENA

Several trends are encouraging some U.S. colleges and universities to offer academic programs outside the United States to non-U.S. students.

The Hunt for Revenue, Prestige and Quality

Revenue generation, prestige, and academic quality - which are potentially, but not necessarily, complementary - are three drivers that may encourage U.S. institutions to engage in cross-border programs. Cuts in state funding for higher education, rising costs, the falling value of endowments, and pressures to limit tuition increases have caused institutions to become more entrepreneurial and more reliant on private sources (Geiger, 2002). Because cutbacks are difficult to make in labour-intensive and highly consultative organizations such as higher education, most institutions have a poor track record of making the deep targeted cutbacks necessary to free needed resources (Eckel, 2003a). They tend to look not to reducing expenses but to expanding revenue. As a result of financial pressures, institutions have intensified their market and market-like behaviour (Slaughter & Leslie, 1997).[39] Entrepreneurial activities that draw from multiple revenue streams become indispensable in a highly competitive postsecondary environment (Hearn, 2003). This drive for revenue has become a major preoccupation of institutional leaders, with some arguing too much so (White & Hauck, 2000).

In addition to revenue, institutions seek prestige (Fairweather, 1988), which participation in the emerging world-wide higher education marketplace can bring. Several scholars (Marginson & Considine, 2000; Slaughter and Leslie, 1997) argue that status is a more important consideration in institutional decision making than money. By enhancing prestige institutions open future possibilities. Positioning themselves favourably in fast-changing education markets may lead colleges and universities to pursue an international entrepreneurial activity that is unlikely to generate immediate financial gain. Often, however, prestige is ultimately related to revenues. Reputation brings public awareness and recognition, opportunities and associations that are not available to lower status institutions (Ehrenberg, 2000).

Finally, working internationally also can enhance quality. Involving faculty and students from abroad can expand expertise, promote adherence to international standards of excellence, and act as an important counterweight to parochialism (de Wit, 2002). Internationalizing teaching and learning offers opportunities for institutions to strengthen liberal learning, helps prepare students for future careers, and enhances students' abilities in an increasingly multi-cultural and globally connected environment (Green, 2003). Many faculty see internationalization as important to broadening students' frames of reference and their range of skills, knowledge and attitudes, and adding vibrancy and relevance to teaching (Green & Olson, 2003). Institutional leaders may further be motivated by the prestige and distinctiveness that a highly internationalized profile brings to an institution.

The Potential of the Curriculum

Most entrepreneurial activities in U.S. higher education focus on either research/ knowledge production or technology transfer. In both cases, the primary resource (or 'capital') is derived from knowledge production (see Ekzkowitz et al., 1998; Fairweather, 1988; Powell & Owen-Smith, 1998). However, to a lesser extent, institutions are beginning to consider their curriculum, courses, and instructional capabilities as capital (Eckel, 2003b), a shift from knowledge production to knowledge dissemination and application as the basis for entrepreneurial activity. In the past, some institutions have found modest ways to tap their curricula to generate income beyond traditional tuition and fees through special programs: for

[39] Slaughter and Leslie (1997, p. 11) make a distinction between market and market-like behavior. Market behavior is the for-profit activity of institutions. Market-like behavior refers to institutional and faculty competition for moneys regardless of its source.

example, Middlebury College's summer foreign language program; American University's semester in Washington D.C. Many others, especially community colleges, provide education and training to corporate partners' employees on a contract basis, sometimes on site and other times via distance learning. For the most part international strategies that capitalize entrepreneurially on the curriculum have been limited to language institutes and intensive summer programs for foreign nationals or graduate programs for foreign nationals. However, emerging trends in cross-border education mean that interest in taking curricula abroad might grow.

Important facilitators of cross-border education are advances in technology and the increasingly widespread use of English as a teaching medium. These two factors enable institutions to reach to a greater number of students, both domestic and international, and increase the potential tuition revenue. Technology can help institutions to transcend the physical time and space constraints of a traditional classroom to reach students in remote locations. Its promise and potential flexibility are appealing to institutional leaders, as is indicated by the rising number of institutions offering individual courses as well as degree programs online (Newman & Couturier, 2001). Instruction in English language is also increasing, particularly in countries that are 'small language' nations such as the Netherlands, Sweden, Norway; and in places that have made a concerned effort to recruit international students. These developments will increase the market for education abroad and make programs from foreign partners that much more attractive.

On a much smaller scale, changes in the process of curriculum development and the nature of faculty work also may facilitate programs abroad or the establishment of campuses in other countries. An emerging practice at many for-profit institutions is to separate the components of a traditional faculty role - course design, text selection, student advising, course delivery, and assessment - and assign them different to different groups or individuals (Kor de Alva, 2003). For instance, at the University of Phoenix an instructional team designs each course, and a second group of instructors delivers it at different sites with a high degree of consistency (Kinser, 2001). The same approach has been used for many years by the non-profit British Open University and by the University of Maryland, University College. This alternative model allows for courses to be taught in multiple locations and maximizes the strengths of people to either design or deliver courses. It creates, as the former president of the University of Phoenix notes, a specialization of function in which no one is expected to be a jack of all trades (Klor de Alva, 2003). Its proponents argue that this approach adds flexibility, streamlines instructional activity, and creates efficiencies.

The Growing Global Student Market

The expanded flexibilities and efficiencies for delivery are driven by growing global demand for higher education. Analysts predict that the worldwide market for higher education in 2025 will approximately double to 160 million students, 45 million of whom will study online (Moe, 2002 as cited by Couturier, 2003). Most developing countries are struggling to meet growing demand for higher education and to increase the percentage of high school graduates entering universities (World Bank, 2000). OECD countries enrolled 1.54 million students in 2001, approximately double the number of 20 years earlier (OECD Secretariat, 2003a).

Government and trans-governmental agencies also are contributing to the robustness of the global student market (van der Wende, 2002). The European Union has made student mobility a priority through such programs as ERASMUS and SOCRATES, and now is using the reforms spurred by the Bologna process to make European higher education more attractive to students from other regions. The adoption of a credit system and a common degree structure of 3 + 2 (three years for the first degree plus two years for a Master's) is intended to make education more flexible and ease mobility. It might also facilitate the development of new programs in Europe by non-European institutions and providers.

The policies of developing countries are also reshaping the global market. Because their higher education systems cannot accommodate projected demand, countries such as Egypt, Chile, Uruguay, Oman, India, China, Hong Kong, Singapore and Malaysia are developing policies to encourage foreign providers, (Couturier, 2003). As part of World Trade Organization (WTO) negotiations of education as a service under GATS, countries such as Congo, Jamaica and Sierra Leone are actively encouraging the entry of foreign institutions and other providers by making unconditional commitments, which means that they place no trade limitations on foreign providers seeking to operate in the country (Knight, 2002).

Although the number of potential students is increasing, so is competition for these students. Australia has tripled its number of foreign students since 1990, and increased numbers more than thirteen-fold since 1980. United Kingdom institutions have increased the number of foreign students three-fold since 1990 and four-fold since 1980 (OECD Secretariat, 2003b). The U.S. has not kept the pace (Schneider, 2000). According to NASFA, the association of international educators, there are currently fewer international students in the U.S. than prior to September 11, 2001); in part because other countries are developing their post-secondary capacities and in part because of 'complacency on the part of the United States' (2006, p. 4). Australian universities have responded to student demand with concerted efforts, predominately in Asia, by engaging in multiple strategies. One is 'offshore education,' whereby students enrol at Australian institutions but study outside the country at a foreign partner's campus. In 2000, approximately one-third of the estimated 100,000 international students enrolled in Australian higher education were located outside of Australia (Davis et al., 2000).

OFFERING AMERICAN DEGREES ABROAD

The most common way for U.S. institutions to participate in the global student marketplace is by enrolling international students in campus-based programs in the U.S. In 2004-05, approximately 565,000 students from abroad studied at American campuses (Chin, 2005). But given the expense associated with studying in U.S. institutions, increased obstacles in obtaining visas, and growing competition from other nations, U.S. institutions may increasingly turn to bringing education to the student as an alternative to bringing students to U.S. campuses. In some instances, as noted, technologically mediated distance education creates a world of new possibilities in this arena. However, other institutions are physically traversing the distances. Some are well-known but most such activity remains uncharted. Webster University, headquartered in Missouri, has six campuses abroad (London, Leiden, Vienna, Geneva, Thailand, Shanghai). Temple University has a campus in Japan, and the University of Chicago Business School operates campuses in Singapore and Barcelona. Kean University in New Jersey is opening a 'full-scale' campus in the Chinese province of Zhejiang. Qatar has invited numerous institutions to offer degree programs. Texas A & M has opened an undergraduate engineering school there, joining Cornell University, which provides a branch of its medical school and Virginia Commonwealth University, which operates a program in fashion and interior design (Mangan, 2003). Georgetown University is planning a program in foreign service as well. Rochester Institute of Technology (RIT) manages the American University of Kosovo and the American College of Management and Technology in Croatia (Gardner, 2003) and the U.S. Business School in Prague. Numerous institutions offer particular programs abroad, such as an MBA degree.

A different strategy to tap the global student marketplace, rather than going it alone or bringing international students to campus, is via partnerships with foreign universities, what we call *curricular joint ventures* (CJVs) (Eckel et al., 2003). This allow institutions to together create and deliver new programs without the risk or expense (and without some of the headaches) of solo efforts. For example, the Singapore-MIT Alliance (SMA) is a global partnership of three research universities: the Massachusetts Institute of Technology (MIT), the National University of Singapore, and Nanyang Technological University also in Singapore. The SMA offers joint graduate education and fosters research collaboration in five areas: advanced materials for micro- and nano-systems, chemical and pharmaceutical engineering, computational engineering, computation and systems biology, and manufacturing systems and technology. Students earn Master's of Science and PhD degrees and certificates. OneMBA is another case, an integrated 21-month international executive MBA program. Partners include the following institutions: Kenan-Flagler Business School of the University of North Carolina, Chapel Hill; Rotterdam School of Management (RSM) of Erasmus University Rotterdam; Chinese University of Hong Kong (CUHK); Monterrey Tech Graduate School of Business Administration and Leadership (EGADE-ITESM) in Mexico; and Brazil's Escola de Administração de Empresas de São Paulo (EAESP), the largest of the schools that make up the Fundação Getulio Vargas (FGV).

AN EMERGING CROSS-BORDER DEGREE FRAMEWORK

Beyond anecdotal understanding, we know little about the extent or types of U.S. degree programs abroad. No comprehensive listing exist. And correspondingly, no systematic understanding exists of the range of approaches being used to provide non-U.S students with U.S. degrees elsewhere. To begin to capture this dynamic trend, the American Council on Education (ACE) proposes the following framework. Its intention is to identify the range of strategies and distinguishing characteristics that differentiate the means through which U.S. institutions (as well as those from other countries) provide their education abroad.

1. Location: In what country is the program offered?
2. Extensiveness of activity. Is the degree program a
    a. Stand-alone comprehensive campus?
    b. Set of degree programs?
    c. Single degree program?
3. Partners: Is the program offered by a sole provider or in conjunction with one or more providers?
    a. If collaborative, are the partners accredited colleges or universities?
    b. If collaborative, are other types of partners (corporate, government, non-profit) involved?
4. Legal Status: Is the program/campus recognized as in the receiving country as
    a. Private, non-profit
    b. Private, for-profit (propriety)
    c. Public
5. Accreditation Status: Is the program accredited?
    a. In the home country?
    b. In the receiving country?
6. Program Level: Is the program offered at:
    a. Pre-collegiate program
    b. Undergraduate/First-Degree Level
    c. Graduate
7. Degrees Awarded: What degree is awarded?
    a. Associates
    b. Bachelors
    c. Masters
    d. Doctorate
    e. Professional
8. Field of Study: In what field of study?
    a. Business
    b. Computer Science
    c. Education
    d. Engineering
    e. Health
    f. Humanities/Fine Arts
    g. Law
    h. Liberal Arts
    i. Medicine

   j. Natural Sciences
   k. Religion/Theology
   l. Social Sciences
9. Curriculum Content: To what extent is the curriculum developed specifically for this site (versus imported from the home institution)?
   a. Completely developed
   b. Completely imported
   c. A mix of imported and especially developed
10. Faculty/Academic Staff: Are the academic staff employed by the receiving country (off-shore) site or are they members of the home institution's faculty/academic staff?
   a. All home (U.S.) institution
   b. All non-U.S. program
   c. Mixture
11. Students:
   a. How many students are enrolled?
   b. What are their countries of origin?
12. Facilities: Does the parent (U.S. institution) own, lease or borrow (via partnership) the site's physical facilities?

The answers to these questions can provide an outline of the various approaches used by institutions in offering degree programs abroad. The assumption is that particular patterns will emerge that suggest particular courses of options for institutions to follow.

THE UNITED STATES ON THE WORLD STAGE

It appears that although degree programs offered by U.S. institutions abroad are not presently a widespread activity, they are steadily increasing. Many signs suggest that more, not fewer, American institutions will look to deliver their academic programs abroad. If entrepreneurial activity and/or a recognized international presence are important for U.S. colleges and universities, technology and changing faculty roles facilitate engagement abroad, and international opportunities and international student markets continue to grow, why is U.S. engagement with other institutions to deliver academic programs seemingly limited? Many possible explanations exist. First, international programs are not trivial undertakings. There are many hurdles, including recruiting students who might not be attracted simply by an international (or American) brand name (Green et al., 2002), negotiating quality assurance mechanisms, setting market-appropriate tuition and fees, and gaining approval to offer degrees. International degree programs also require a significant investment of time and money. Most programs to date seem to have been the province of those institutions that can make such investments, or those institutions on the financial margin willing to take substantial risks. Second, in tight financial times, it can be difficult to convince board members, faculty decision makers and others that the use of scarce institutional resources of money and time for international ventures is more important than attending to students' at-home needs. Third, returns on investment are uncertain and most likely not immediate. Programs abroad can have tremendous start-up costs, such as travel, IT infrastructure, instruction, and marketing. Establishing presence in an emerging market carries financial risks. Finally, and importantly, most U.S. institutions look to international activities that will benefit their own students and faculty, enriching their learning environment while serving the needs of their partners.[40]

Does the current pattern of cross-border engagement by some U.S. institutions only suggest the global marketplace is the playground of the few? We can only hypothesize. It appears that only some specific types of U.S. institutions are taking the plunge and offering programs outside the nation's borders. First, large research universities may increasingly operate cross-border because they see their mission as that of a 'world-class university' or because these strategies help garner prestige. When they join forces with other prestigious institutions this helps to enhance reputation and adds intellectual firepower. Second, a very small number of institutions define their institutional identity and image as internationally focused. By operating abroad they are pursing a specific market niche. The third group of institutions appears to be focused on revenue generation first of all. Many for-profit institutions may fall into this category. It is possible that other institutions will enter the international arena when the investments costs lower and returns are more certain and/or when they can identify willing partners guaranteeing access to growing markets or technologies. However, there will always be an element of risk. It is unclear whether significant numbers of institutions will want to incur the costs and take the risks, given the types (and uncertainty) of the rewards.

CONCLUSION: A CHECKLIST OF GOOD PRACTICE

In 2004 the American Council on Education (ACE), the Association of Universities and Colleges of Canada (AUCC), the Council for Higher Education Accreditation (CHEA), and the International Association of Universities (IAU) issued a statement, *Sharing quality higher education across borders: A statement on behalf of higher education institution's worldwide* (ACE et al., 2004). That statement set out a series of principles to guide provision of cross-border education and made a set of recommendations for higher education institutions and other providers and governments to follow. More than 30 higher education associations worldwide have endorsed the statement. The principles include the following:

- Cross-border higher education should strive to contribute to the broader economic, social and cultural well-being of communities.
- While cross-border education can flow in many different directions and takes place in a variety of contexts, it should strengthen developing countries' higher education capacity in order to promote global equity.

[40] The latter motivation is not, however, universally applicable to U.S. institutions. Some are deeply motivated by entrepreneurial rewards. See Kirp & Roberts, 2002; Slaughter & Leslie, 1997.

- In addition to providing disciplinary and professional expertise, cross-border higher education should strive to instill in learners the critical thinking that underpins responsible citizenship at the local, national and global levels.
- Cross-border higher education should be accessible not only to students who can afford to pay, but also to qualified students with financial need.
- Cross-border higher education should meet the same high standards of academic and organizational quality no matter where it is delivered.
- Cross-border higher education should be accountable to the public, students and governments.
- Cross-border higher education should expand the opportunities for international mobility of faculty, researchers and students.
- Higher education institutions and other providers of cross-border higher education should provide clear and full information to students and external stakeholders about the education they provide.

As a follow-up to that Statement, the same four organizations released a checklist for good practice framing a series of questions for institutional leaders to use in designing and assessing their cross-border educational initiatives and putting the principles of the 2004 statement into practice (ACE et al., 2005). The checklist focuses on eight key areas and asks the following questions:

Contribution to broader public good:

- Does the institution's statement of mission and goals include a commitment to the social, cultural, and economic well-being of communities?
- Does the institution's statement of mission and goals include a commitment to international education, outreach, and linguistic and cultural diversity?
- Are the institution's cross-border programs culturally sensitive in approach and content? How does the institution evaluate this?

Capacity building

- In designing its cross-border educational initiatives, has the institution consulted with local institutions regarding how its programs might enhance the overall capacity of higher education? For example, is there an opportunity to cooperate, when appropriate, with local institutions?
- Do the institution's cross-border programs encourage collaboration with the host country institutions and expand the opportunities for international mobility of faculty, researchers, and students?

Relevance

- Do the institution's mission, goals, or assessment measures for student learning include a commitment to the importance of critical thinking that underpins responsible citizenship at the local, national, and global levels? Has this commitment been integrated into the institution's cross-border initiative?

- Are the degrees and other credentials granted by the institution in its cross-border programs recognized for purposes of employment in the countries in which the students are enrolled?

Accessibility

- Does the institution enhance access to its programs and courses by offering support to qualified students with financial need from the host country or other countries?

Quality

- Does the institution have in place a process of ongoing quality review, feedback, and improvement that relies on faculty expertise and incorporates the views of students? Has the institution taken steps to promote the application of this process to its educational initiatives abroad?
- Does the institution apply the same quality assurance principles, policies, and practices—and standards of academic and organizational quality—no matter where its programs are delivered?
- Does the institution employ the same standards and procedures in appointing and evaluating faculty members, wherever the instruction is offered?
- When instruction is provided primarily or entirely over the internet or by other electronic means to students in the host country, does the institution also provide appropriate technical support for students and faculty, access to library resources, advising, and other student services?
- Does the institution provide advising and orientation support for students, wherever they are enrolled?
- Does the institution provide adequate administrative support for its programs and activities, wherever they are offered?

Accountability

- Has the institution obtained the proper authorization to operate and offer programs as a higher education institution from government or other competent bodies in the home and host countries?
- Does the institution engage and cooperate with its associations and governmental and nongovernmental bodies to develop effective quality assurance principles and practices for cross-border educational initiatives and to improve the international exchange of information and cooperation on quality assurance and recognition issues?

Transparency

- Does the institution provide accurate and user-friendly information to the public, students, and governments in a proactive manner, particularly with respect to the institution's legal status, award-granting authority, course/program offerings, quality assurance mechanisms, and other relevant facts as suggested by codes of good practice?

- Does the institution have clearly stated policies and criteria for the admission of students, expectations of student performance, total cost of enrollment, and availability of financial aid?
- Does the institution maintain adequate, appropriate, and clearly stated procedures for student appeals of grades and other academic policies and decisions?

Commitment to high-quality higher education across borders

- Has the declaration, *Sharing quality higher education across borders: A statement on behalf of higher education institutions worldwide*, or other codes of good practice in international education and cross-border education provision been disseminated to interested faculty, staff, and students in the institution?
- Does the institution adhere to specifically identified codes of good practice in higher education and cross-border education provision?

The intent of this checklist is to help campus leaders make focused decisions to safeguard the cultural, social and economic contributions of higher education, protect the interests of students and facilitate their mobility, and strengthen the capacity of developing countries to increase access to high quality education. Without these priorities, higher education all too easily becomes simply another category of trade to be governed by the ever-widening GATS framework. It is the adherence to these principles, practices and values, that will help distinguish worldwide higher education from other market-driven goods and services.

REFERENCES

American Council on Education (ACE), Association of Universities and Colleges of Canada (AUCC), Council for Higher Education Accreditation (CHEA), & International Association of Universities (IAU) (2004). *Sharing quality higher education across borders: A statement on behalf of higher education institution's worldwide*. Accessed at: http://www.unesco.org/iau/p_statements/index.html

American Council on Education (ACE), Association of Universities and Colleges of Canada (AUCC), Council for Higher Education Accreditation (CHEA), & International Association of Universities (IAU) (2005). Questions for institutional leaders to use in designing and assessing their cross-border educational initiatives. Accessed at: http://www.acenet.edu/AM/Template.cfm?Section=Search&Template=/ CM/ContentDisplay.cfm&ContentFileID=1643

Chin, H.-K. (2005). *Open Doors 2005*. New York: Institute for International Education.

Couturier, L. K. (2003). *Globalizing with a conscience: The implications of privatization in higher education*. Paper presented at the conference 'Markets, profits and the future of higher eucation. May 1-2. New York: Teachers College, Columbia University

Davis, D., Olsen, A., & Bohm, A. (2000). *Transnational education providers, partners and policy: Challenges for Australian institutions offshore*. Brisbane: IDP Education Australia.

de Wit, H. (2002). *Internationalisation of higher education in the United States of America and Europe*. Westport, CN: Greenwood Press.

Eckel, P. D. (2003a). *Changing course: Making the hard decisions to eliminate academic programs.* Westport, CT: Praeger.

Eckel, P. D. (2003b). Capitalizing on the curriculum: The challenges of curricular joint ventures. *American Behavioral Scientist,* 46(7), 865-882.

Eckel, P. D., Affolter-Caine, B., & Green, M. (2002). *New times, new strategies: Curricular joint ventures.* Washington DC: American Council on Education.

Ehrenberg, R. G. (2000). *Tuition rising: Why college costs so much.* Cambridge, MA: Harvard University Press.

Ekzkowitz, H., Webster, A., & Healey, P. (1998). *Capitalizing knowledge: New intersections of industry and academia.* Albany, NY: State University of New York Press.

Fairweather, J. S. (1988). *Entrepreneurship and higher education.* ASHE-ERIC Higher Education Report No. 6. Washington DC: The George Washington University.

Gardner, E. (2003). Going global: American colleges and universities head to distance lands, and approach the challenge in markedly different ways. *University Business.*

Geiger, R. L. (2002). The American university at the beginning of the twenty-first century: Signposts on the path to privatization. In *Trends in American and German higher education.* Cambridge MA: American Academy of Arts and Sciences, 33-84.

Green, M. (2003) The internationalised campus: A strategic approach. *International Educator,* 12(1), 13-21.

Green, M., Eckel, P., & Barblan, A. (2001). *The brave new (and smaller) world: A transatlantic view.* Changing Enterprises Occasional Paper No. 1. Washington DC: American Council on Education.

Green, M. & Olson, C. (2003). *Internationalizing the campus: A user's guide.* Washington DC: American Council on Education.

Hearn, J. C. (2003). *Diversifying campus revenue streams: Opportunities and risks.* Informed Practice Series. Washington DC: American Council on Education.

Kinser, K. (2001). *Faculty at the University of Phoenix: A study of non-traditional roles.* Paper presented at The Association for the Study of Higher Education annual meeting, 17 November. Richmond, VA.

Kirp, D. L. & Roberts, R. S. (2002). Mr. Jefferson's 'private' college: The University of Virginia's business school secedes,. *Occasional Paper No. 55,* National Centre for the Study of Privatization in Education. New York: Teachers College, Columbia University.

Knight, J. (2003a). Updating the definition of internationalization. *International Higher Education, 33,* 2-3.

Knight, J. (2003b). Trade talk - à la four modes. *International Higher Education, 31,* 1-2.

Knight, J. (2002). *Trade in higher education services: The iImplications of GATS.* Report 10 of The Observatory on Borderless Higher Education (OBHE). London: OBHE.

Kor de Alva, J. (2003). *Higher education productivity: Closing attainment gaps and the maintenance of a global competitive edge.* Presentation to the National Governors Association Centre for Best Practices, September. San Francisco.

Mangan, K. S. (2003). Texas A&M to open program in Qatar. *The Chronicle of Higher Education,* 6 June, A42.

Marginson, S. & Considine, M. (2000). *The enterprise university: Power, governance and reinvention in Australia.* New York: Cambridge University Press.

NASFA, Association of International Educators (2006). *Restoring U.S. competitiveness for international students and scholars.* Washington DC: NASFA. Accessed on 20 June 2006 at: http://www.nafsa.org/_/Document/_/restoring_u.s.pdf

Newman, F., & Couturier, L. (2001). *The new competitive arena: Market forces invade the academy.* Working Paper, The Futures Project: Policy for Higher Education in a Changing World. Providence, RI: Brown University.

OECD Secretariat (2003a). *Key developments and policy rationales in cross-border post-secondary education.* OECD/Norway Forum on Trade in Educational Services. OECD: Paris.

OECD Secretariat, (2003b). *Cross-border education: An overview*. OECD/Norway Forum on Trade in Educational Services. OECD: Paris

Powell, W. W. & Owen-Smith, J. (1998). Universities as creators and retailers of intellectual property: Life-sciences research and commercial development. In B. A. Weisbrod (Ed.) *To profit or not to profit: The commercial transformation of the nonprofit sector,* 169-194. New York: Cambridge University Press.

Schneider, M. (2000). *Others' open doors - How other nations attract international students: Implications for U.S. educational exchange*. Available at: http://opendoors.iienetwork.org/

Singapore-MIT Alliance (SMA) (2006). Partnership of the Massachusetts Institute of Technology (MIT), National University of Singapore, and Nanyang Technological University. Website accessed at: http://web.mit.edu/sma/

Singh, M. (2002). International quality assurance, ethics, and the market: A view from developing countries. In *Globalization and the market in higher education: Quality, accreditation and qualifications*, 171-190). Paris: UNESCO.

Slaughter, S. & Leslie, L. L. (1997). *Academic capitalism: Politics, policies, and the entrepreneurial university*. Baltimore: Johns Hopkins University Press.

Van Damme, D. (2002).Higher education in the age of globalization. In *Globalization and the market in higher education: Quality, accreditation and qualifications*, 21-34. Paris: UNESCO.

Van de Wende, M. (2002). Internationalisation policies: About new trends and contrasting paradigms. *Higher Education Policy*, 14, 249-259.

World Bank, (2000). *Higher education in developing countries: Peril and promise*. Report by The Task Force on Higher Education and Society. Washington, D.C.: World Bank.

Peter D. Eckel & Kate Berniaz
Centre for Effective Leadership
American Council on Education
Washington, United States of America

Madeleine F. Green
Centre for International Initiatives
American Council on Education
Washington, United States of America

BRIAN PUSSER, WALDEMIRO VÉLEZ-CARDONA & GLORIA DÁVILA-CASASNOVAS

7. PELL GRANTS AND STUDENT CHOICE IN PUERTO RICO:

Subsidies and Postsecondary Transformation[41]

INTRODUCTION

A prominent development in public policy proposals for postsecondary education over the past two decades has been an increased demand for approaches to university organization and finance based on market models and free-market entrepreneurial behavior (Marginson, 1997; Winston, 1999; Newman, 2001; Pusser & Doane, 2001; Ehrenberg, 2000; Kirp, 2003; Geiger, 2004; Pusser, 2006). National political leaders, state governors, legislators, policy analysts, and a variety of interest groups have increasingly proposed that universities reorganize to privilege student choice, consumerism, competition, privatization, and deregulation as keys to strategic planning and adaptation in the twenty-first century (Collis, 1999; Duderstadt, 1999; Kirp, 2003; Pusser & Turner, 2004; Pusser, 2006).

The belief that the efficient provision of postsecondary education necessitates the use of market principles and behaviours is now a key factor in global higher education debates over how higher education subsidies should be allocated (Marginson & Considine, 2000; Johnstone, 2004; Jongbloed, 2004; Slaughter & Rhoades, 2004). These contests have been shaped by two persistent strands of political pressure, each of which serves as part of a broader restructuring of public institutions (Slaughter & Rhoades, 2004; Marginson, 2004; Pusser, 2006b). The first strand is the continued retreat by the State from collective public funding for many public institutions, including higher education, as part of a long-term trend away from a reliance on tax revenue as the primary support of public institutions (Breneman, 2005; Breneman, Pusser & Turner, 2006; Vélez, 2002). One manifestation of this trend is a significant decline in the per capita public funding allocations for higher education in most states of the United States and in a wide range of countries (Newman et al., 2004 Marginson & Considine, 2000). Another is the re-conceptualization of higher education as a private good (Pusser, 2006a).

The second political project driving public sector restructuring has been the effort by many political leaders and interest groups to conceptualize nearly all public sector institutions, including colleges and universities as inefficient, lacking

[41] Portions of this paper appear in Pusser, B. (in press) From ideology to policy: The evolution of choice programs in higher education, La Revista de El Centro de Estudios y Documentacion sobre la Educacion Superior Puertorriquena [Review of the Centre for the Study of Puerto Rican Higher Education] San Juan: Puerto Rico.

Simon Marginson (ed.), Prospects of Higher Education, 155–169.

accountability and suffering from a lack of competition. This conceptualization has achieved taken-for-granted status in the United States despite a dearth of empirical evidence. In turn, the belief that markets and individual choice are superior to State provision in the public sector has achieved a similar taken-for-granted status (Vélez, 2006). Amaryta Sen suggests, in a work generally supportive of market mechanisms, 'the virtues of the market mechanism are now standardly assumed to be so pervasive that qualifications seem unimportant. Any pointer to the defects of the market mechanism appears to be, in the present mood, strangely old-fashioned and contrary to the contemporary culture (like playing an old 78 rpm record with music from the 1920s)' (Sen, 2000, p. 11). Nonetheless, a growing body of research in higher education has begun to question whether choice models are appropriately applied to either elementary-secondary or higher education (Levin, 2001; Doane & Pusser, 2001; Marginson, 2004; Slaughter & Rhoades, 2004).

Interest group and legislative efforts to restructure higher education over the past decade (Levin, 2001; Vélez, 2006) have promoted the premise that consumer choice and institutional competition will lead to greater efficiency and accountability (Friedman, 1962, 1980; Marginson, 2004). A number of U.S. states have adopted language that initially appeared as part of the education policy debate over the creation of vouchers in the K-12 educational system. An analysis of state higher education policy documents since the mid 1980s found myriad calls for increased choice, for the conceptualization of students and families as consumers at the centre of the postsecondary project, for reduced regulation and for increased competition for State resources (Pusser & Doane, 2006). These policy declarations reflect the central tenets that have emerged from neo-liberal economic restructuring proposals advanced in many parts of the world Colclough notes: 'If there were to be a single statement which symbolized the fundamental message of neo-liberalism, it be would be this: in settling matters of resource allocation, imperfect markets are better than imperfect states' (Colclough, 1991, p. 7).

Public and Private Goods in Higher Education

The market discourse in policy debates over higher education has been accompanied by an equally important shift in the discourse shaping perceptions of the public and private goods generated by the sector. That debate, essentially a continuation of one dating back to the early arguments about human capital and education (Becker, 1964), suggests that higher education is a private good and that as such, much of the cost should be borne by individuals not public (Marginson, 2004; Pusser, 2006). This shift also questions the traditional organization of higher education in the United States as a collaborative project between the public, postsecondary institutions and the professionals labouring within those institutions. The concept of higher education as a private good moves policy makers away from collaborative provision and calls for structures that support private initiative and attainment. This in turn fuels demands for higher student tuition and a policy preference for portable financial aid over direct allocations to institutions.

Over the past decade, fundamental attributes of market approaches have been moving to the centre of global higher education policy recommendations by such influential organizations as the OECD (Lingard & Rizvi, 1998) and in the United States by the National Governor's Association (NGA) and individual state legislative bodies and coordinating boards (Kelly, 2001). Most prominent among these market attributes is the notion that increased consumer choice will improve the provision of public higher education. A report commissioned as part of The NGA's initiative, Influencing the future of higher education (2001), notes that:

> Savvy states in the twenty-first century will focus on postsecondary customers: the learner, the employer, and the public who supports educational opportunities. In competitive states, resources will increasingly flow to the learner, and state regulatory policies will ease to encourage institutional flexibility (NGA, 2001, p. 3).

A special commission on higher education in Texas concluded earlier this decade that the state should develop a funding system for public higher education driven by consumer choice. The commission suggested a funding model that 'emphasizes grants to students over direct appropriations to institutions' (Special Commission on 21st Century Colleges and Universities, 2001, p.12). A variation on this plan has been adopted in Colorado and is being considered in several other states.

These initiatives for restructuring the organization and finance of public higher education have major implications. A change from state block funding of institutions to direct funding of learners would constitute a new direction in the public financing of U.S. higher education, one of the more significant shifts since the Higher Education Act (HEA) of 1965 and the HEA Amendments of 1972 (Breneman, 1992; Pusser & Doane, 2001). But a central question remains: does this shift from the financing of institutions to the financing of individuals affect the institutional array, student choice of institutions and the type of educational opportunity available to students? In turn, would these changes also alter the balance of public and private goods (Pusser, 2006) produced by higher education?

The chapter. The chapter applies empirical data to the question of how market competition and student choice shape institutional success and the production of public goods. We evaluate data from the volution of the Puerto Rican system of higher education to gain insight into the impact of a rapid shift in the form and amount of student subsidies on institutional formation, enrolments and student choice. The Pell Grant program in Puerto Rico offers a model of a well-funded voucher system implemented in a relatively undeveloped postsecondary system.

BACKGROUND

The Pell Grant

The Pell grant is named after Senator Claiborne Pell of Rhode Island, one of the principal architects of the federal student aid programs adopted in the mid-1960s and early 1970s. Since their creation in 1972, Pell grants have become the largest need-based federal grant system (awarded independent of academic achievement) in the United States (Suspitsin, 2002). Some 95 percent of Pell grants go to students with annual family income below $40,000. In fiscal year 2005, Pell grant awards to students ranged from $400 to $4050, and the total amount awarded to 5.3 million recipients was just over $12 billion (Mercer, 2005). According to a recent assessment conducted by the U.S. Office of Management and Budget:

> The Pell Grant program compares very favourably with other programs with similar purpose and goals; its voucher-like structure maximizes student choice and simplifies the delivery of funds, while studies have consistently found that Pell Grants are the single most effective tool in increasing low-income access to higher education (OMB, 2005).

The Case of Puerto Rico

The Commonwealth of Puerto Rico is located on a set of islands about 1,000 miles southeast of Miami, Florida, with about four million people. Residents of Puerto Rico hold United States Citizenship, are subject to U.S. federal laws and are eligible to receive various forms of federal aid (CRS Report for Congress, 2005). Puerto Rico has a per capita GDP in 2005 of approximately $18,000 as compared to a per capita GDP for the United States in that same year of $42,000. The Commonwealth has long dealt with sustained high rates of unemployment. Since the early 1970s the rate of unemployment has remained above 10 percent and occasionally reached levels twice that high (DTRH, Selected Years).

The higher education system in Puerto Rico was relatively undeveloped until the latter half of the twentieth century. While the University of Puerto Rico was founded in 1903, by the early 1950s there were only five universities on the islands, graduating about 5,000 students annually (Roucek, 1953).[42] Puerto Rican world war 2 veteran were eligible for support under the Serviceman's Readjustment Act (the GI Bill) but their enrolments did not significantly shape the postsecondary system (Roucek, 1953). Along with raising the educational and skill levels of the population and stimulating national economic development, an important contribution of the postsecondary system has been to offer an alternative to long-term unemployment for young Puerto Ricans (Duncan, 1981).

Hence unlike the continental United States, with a highly developed system of public, private and for-profit postsecondary institutions at the time of the passage of HEA, Puerto Rico's system was less fully developed and less funded. Because Puerto Rico was included in the financial aid provisions of HEA 1972, the degree and form of higher education subsidies available to Puerto Rico's postsecondary system changed dramatically. Since the beginning of Pell Grants, Puerto Rican postsecondary students have been very dependent on them and are now major beneficiaries of US-based financial aid. Undergraduate enrolment in Higher

[42] The University of Puerto Rico, Rio Piedras, classified as a Carnegie Research University/High Research Activity institution was the only research university in Puerto Rico in 2005-2006.

Education Institutions (HEIs) in Puerto Rico was just under 200,000 in 2004-2005, a little over 1 percent of total postsecondary enrolments in the United States in FY 2004-2005. But that year almost 5 percent of total U.S. Pell Grant disbursements went to Puerto Rico. Pell Grants represented 69.7 percent of all federal student aid funds received in Puerto Rico for FY 2005-2006, though in the United States Pell grants represented only 12 percent of total student aid dollars in the same year (PRCHE, 2006; U.S. DOE Budget Service, 2007; U.S. DOE Overview, 2007).

Over the past decade the percentage of postsecondary students in Puerto Rico in receipt of Pell grants has fluctuated between 80 and 90 percent. In the United States fewer than 30 percent of postsecondary students are Pell recipients, which in 2001 extended to more than half of the students in only one state (Mississippi, (53.8 percent). In 2004-2005 the average Pell grant in Puerto Rico was $3022, for the entire U.S. it was just under $2500 (U.S. Department of Education, 2006).

Since the origination of the program, Pell Grants have served as the primary form of student aid in Puerto Rico and as a key driver of institutional viability. The Pell Grant has also served as a major stimulus for postsecondary institutional formation and transformation in Puerto Rico. Hence the appearance of portable financial aid in Puerto Rico and the significance of what has effectively become a higher education voucher system for the Island renders it a useful site of study.

Research Questions

Given that the compelling narrative behind many contemporary postsecondary policy initiatives is that greater consumer choice will increase competition, efficiency and the diversity of institutional forms in postsecondary education, we looked at the data from Puerto Rico with four primary research questions in mind:

1. To what degree do new subsidies provided directly to students reconfigure the postsecondary institutional landscape?
2. Do new subsidies directed to students fuel growth in existing institutions?
3. Do such subsidies encourage formation of new postsecondary institutions?
4. If new institutions are founded in response to these subsidies, are they predominantly public or private institutions?

Data sources: Data for this research are from the NCES/IPEDS the Council on Higher Education of Puerto Rico and the Puerto Rican Higher Education Research and Information Centre of the University of Puerto Rico. The data, for the period 1970-2000, include the percentage of the 18-24 year-old population in postsecondary education, the number and percentage in public institutions and the number and percentage in private institutions. Data were also gathered on financial aid received by students for 1970-2000, the number of institutions offering postsecondary education, the number and percentage of public and private institutions and student enrolments in each institutional type.

These data are supplemented by analyses of institutional characteristics, student graduation rates and accreditation reports.

Table 7.1. Number of and enrolment in higher education institutions (HEIs), Puerto Rico

years	*number of institutions*			*enrolment by institutional type*		
	private	*public*	*total*	*private*	*public*	*total*
1970	16	12	28	21,933 (34%)	42,516 (66%)	64,449
1974	20	17	37	42,314 (45%)	52,055 (55%)	94,369
1980	34	19	53	79,234 (59%)	55,919 (41%)	135,153
1990*	48	20	68	99,366 (63%)	57,997 (37%)	157,363
2000	53	19	72	110,997 (60%)	74,018 (40%)	185,015

*One HEI was classified public that year and private in 2000. HEI grant at least Associates of Arts (A.A.) degrees. sources: Puerto Rican Council of Higher Education and IPEDS data.

PATTERNS OF CHANGE

Pell Grant Awards in Puerto Rico

In FY 1974[43], the first year of Pell eligibility in Puerto Rico, the amount of financial aid available to students in Puerto Rico from Commonwealth and local sources was just over $6.0 million. In that same year, Pell grants provided an additional $14.0 million to students in Puerto Rico, essentially tripling the total financial aid delivered in one year (Table 7.1) (US Department of Education, Selected Years).

From FY 1974 to 1980, the first six years of Pell eligibility, while the 18-24 year olds enrolled in higher education rose from 26 to 34 per cent of the age group in the general population, total Pell grants increased by over 900 percent to $135.0 million (US Department of Education, Selected Years; PRCHE, Selected Years).

From 1980-2000, while student enrolments increased by 37 percent, the amount of Pell grants received by undergraduates in Puerto Rico jumped to $448.6 million, an increase of 231 per cent. Pell grants in the year 2000 accounted for about 70 percent of all financial assistance received by undergraduate students in Puerto Rico, while and the percentage of the eighteen- to twenty four year old population enrolled in postsecondary education increased to just under 43 per cent (US Department of Education, Selected Years; PRCHE, Selected Years) (Table 7.2).

Enrolments and Institutional Array

The distribution of enrolments within the Puerto Rican postsecondary system also shifted dramatically. In 1970, 66 percent of postsecondary students were enrolled

[43] Pell eligibility was increased in stages after its initial passage. In 1973/74 it was available to freshmen only. By 1976/77 Pell Grants were available to all undergraduates enrolled at least half-time in Pell eligible institutions.

in public institutions and 34 percent in privates (Table 7.1; PRCHE, 1972). By 2003 only 36 percent of postsecondary students were in public institutions.

Table 7.2. Number of students receiving Pell grants and total grants received, Puerto Rico

years	population total	population 18-24 years	enrolment: total	enrolment: proportion of 18-24s	Pell grants total * (millions, constant dollars)
1970	2,712,033	337,742	64,449	19.08	0
1974	2,881,000	360,125	94,369	26.20	13.372
1980	3,196,520	395,439	135,153	34.18	** 135.509
1990	3,522,037	415,854	157,363	37.84	240.545
2000	3,808,610	430,852	185,015	42.94	448.600

* constant prices ** year 1981-1982
Sources: Puerto Rican Council of Higher Education and IPEDS data

A total of 64 percent were in private postsecondary institutions (Figure 7.1 below). While enrolments for proprietary institutions are not available for 1974, document analysis suggests these have risen rapidly in the past three decades.

The infusion of portable student financial aid to Puerto Rico that began in 1974 had a profound impact on the distribution of institutions and student enrolments. In 1974 there were 37 higher education institutions accredited by the Council on Higher Education of Puerto Rico (CHE): 20 private (54 per cent) and 17 public (46 per cent). These 37 institutions enrolled 94,369 undergraduate students, with 45 per cent of enrolment in private institutions and 55 per cent public (see Table 7.1). Six years later, in FY 1980 there were 53 institutions accredited by the Council on Higher Education: 34 private (64 per cent) and 19 public (36 per cent). In FY 1980 there were 135,153 undergraduates enrolled in Puerto Rico; 79,234 enrolled in private institutions (59 per cent) and 55,919 in public institutions (41 per cent).

The trend continued over the next twenty years. By FY 2000, there were 72 institutions accredited by the Council, 53 private (74 per cent) and 19 public (26 per cent). The same year there were 185,015 undergraduate students, 110,997 in private institutions (60 per cent) and 74,018 (40 per cent) in public institutions (Tables 7.1 & 7.2). The most recent data show an increasing shift in favour of private enrolments. In Fall 2003, just under 65 per cent of undergraduate enrolments were in private two- or four-year institutions. About 8 per cent were in for-profit two- or four-year institutions (PRCHE, Selected Years).

For-Profit Institutions in Puerto Rico

As dramatic as the increase in the growth of institutions and enrolments in the Puerto Rican private postsecondary system has been since the introduction of Pell grants, special notice should be made of the growth of for-profit institutions. In 1970 there were no for-profit HEIs accredited or licensed by the Council of Higher Education. For-profits were just pre-university institutions (less than two years).

Neither licensing nor accreditation was required for a for-profit institution to operate in Puerto Rico. The license to operate became compulsory in 1976.

Figure 7.1. Pell grant distribution by institutional type, Puerto Rico compared to rest of United States, 2003

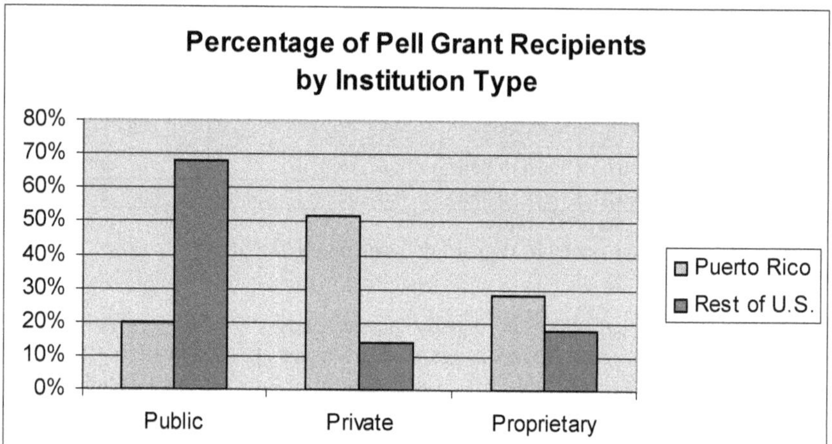

Source: 2004-2005 Federal Pell Grant Program End-of-Year Report, U.S. Department of Education

By 1979-80, there were three for-profit institutions operating six facilities, two-year institutions with a combined enrolment of 3,605 students. That represented 4.5 per cent of total enrolment in private HEIs and 2.7 per cent of the total postsecondary student enrolment on the Island. By 1989-90, there were 11 for-profit facilities: four four-year institutions and seven two-year institutions, with a total enrolment of 5,565 students, 5.6 per cent of the total enrolment in private institutions and 3.5 per cent of the total student enrolment in higher education. In 1999-2000 there were 14 for-profit facilities, five four-year institutions and nine two-year institutions with a total enrolment of 12,055 students, 10.9 per cent of private postsecondary enrolment and 6.6 per cent of all postsecondary enrolment.

Six years later in 2005-2006, the number of for-profit institutions licensed to operate in Puerto Rico had reached 18, of which seven were four-year institutions and 11 two-year institutions. Together they enrolled 16,488 students, which was 12.4 per cent of the total enrolment in private HEIs and 8.0 per cent of enrolment in postsecondary education Puerto Rico (PRCHE, Selected Years; see Table 8.2). In just over twenty-five years, the number of for-profit institutions more than tripled, their share of the private sector enrolment increased 180 per cent, and their share of the total postsecondary enrolment in Puerto Rico rose over 200 per cent.

The literature on for-profit institutions suggests that their recent emergence has significant implications for equity, efficiency and the production of public goods through higher education (Pusser, 2006; Kirp, 2003). The growth of for-profits in Puerto Rico is likely to have significant impacts on all of these dimensions.

The Expansion of Higher Education: A Great Benefit for Puerto Rico?

At the end of last century there was much debate and discussion of the challenges and problems emerging from the rapid development of higher education. There was dissatisfaction with the challenges to planning and development caused by the low proportion of funding provided by the Commonwealth of Puerto Rico. A study sponsored by the Council on Higher Education (PRCHE) in 1997 states:

> The scant appropriation of economic resources for higher education from federal appropriations has reduced the capacity of the government to influence the evolution of the system. What has shaped the system are the regulations and guidelines of federal aid programs. At present we find ourselves at a crossroads between the way in which the higher education system has actually developed and the needs and goals of Puerto Rican society. This dilemma requires the establishment of public policies to rationalize the strategic development of the system, (PRCHE, 1997, p.10).

From 1997-2000 the PRCHE sponsored a series of dialogues and work groups to create proposals for policies to reform higher education in Puerto Rico. This culminated in a strategic planning document entitled *Higher education in Puerto Rico: Toward a vision of the future* (2000). It summed up the situation as follows:

> During the second half of the 20th century, higher education in Puerto Rico experienced an accelerated expansion and diversification. This growth was not sufficiently articulated, so that rather than having a coherent development process, institutions in both the public and private sectors have operated out of their own self-interest. Such influential issues as the social value of education, access policies, increases in the availability of economic resources for education and enrolment growth have evolved during the expansion of higher education without sufficient planning. This expansion has also not always generated programs with the degree of social relevance or quality required to enable HEIs to meet the country's needs (PRCHE, 2000, p.5).

There was concern that the expansion of higher education in Puerto Rico through consumer demand and market competition may have compromised the integrity and credibility of the system:

> In a globalized world that is ever more complex, pluralistic and defined by accelerated change, a diverse system of Higher Education Institutions is a welcome phenomenon. In Puerto Rico the system has diversified, but this has been driven by the market for student enrolments rather than by the need for social and cultural development in the country. A postsecondary system without congruence between the actual practices of the institutions and what

> the institutions represent to society compromises the integrity and credibility of the system. In many cases beliefs about institutional purposes (in Puerto Rico) are not congruent with the actual institutional missions, curricula, the student populations, services, faculty expertise and research capabilities, or degrees granted, (PRCHE, 2000, p.7).

The fact that a market driven by the availability of Pell Grants has been the main force shaping the growth and diversification of higher education has significantly limited the ability of the higher education system to evolve in service of the needs of Puerto Rican society. The inordinate effort given to responding to the demands of the market has resulted in an expansion in which extreme competitiveness between institutions has resulted in a narrow focus, the duplication of offerings, and loss of opportunity (PRCHE, 2000). As Quintero and Irizarry (2006) observe:

> Over the most recent period (2000-2005) there has been a proliferation of new programs without direction or coherence. The expansion of offerings and the increase in enrollments has occurred primarily in the private sector through the creation of new undergraduate and graduate programs. Yet the majority of new associate's and bachelor's programs are duplicate offerings in areas such as business management and teacher preparation. Across the system there is already an excess of such programs despite the employment possibilities they may entail (Quintero e Irizarry, 2006, p.37).

The Market, the University and Economic Development

Over the past two decades the government of the Commonwealth has been trying to position Puerto Rico in a global knowledge-based economy, through strategies designed to attract high technology industries, particularly in biotechnology. There has also been significant effort to foster research and development programs to promote creation of small and mid-sized local high technology industries (SMI's) capable of integrating in clusters with advanced technology sectors in the life sciences, telecommunications and computing. The strategy has been called 'minds at work' to differentiate it from the previous development strategy of 'hands at work' (Dietz, 1987). The new strategy will require a significant contribution from the higher education system. But the consumer-driven growth and expansion of the postsecondary sector has not built the appropriate foundation for it to contribute effectively to the program of economic development the country is undertaking.

Analysis of government reports, and recent research on the Puerto Rican system, suggests that expansion through the use of student vouchers has been a mixed blessing. Coordinated efforts, including State regulation and financial incentives to shape the provision of postsecondary education would probably have resulted in greater benefits, particularly in relation to science and technology capacity. The PRCHE notes that research in science and technology has been concentrated in the three main campuses of the University of Puerto Rico, a public research university, that since the 1960s has been seen as a key site for scientific innovation, research and development programs. Yet despite the growth of institutions and enrolments the Commonwealth has not created a critical mass of science and technology expertise, nor a national research and development system based on higher education (PRCHE, 2004; Ríos, 2005; Iturralde, 2005). According to Stewart:

> In 2002, public universities and colleges in Puerto Rico spent $901 on research and development per student FTE. This was 32.9 per cent of the expenditure per student for public universities and colleges by public colleges and universities in the United States. Private universities and colleges in Puerto Rico spent only $111 per FTE enrolment, about 3 per cent of the amount spent by private universities and colleges in the US. (2006, p. 17).

Stewart also notes that the share of investment in Puerto Rican university research and development relative to such investment in the United States has declined since the introduction of Pell Grants. The ability of the Puerto Rican postsecondary system to generate research and development is directly linked to the health of the public higher education institutions. The University of Puerto Rico currently has sixteen doctoral programs in the basic sciences and engineering. However, an analysis of private universities found only one (Quintero e Irizarry, 2006).

The Emerging Market and Student Success

While there has been a significant increase in enrolments in private postsecondary institutions in Puerto Rico since the adoption of Pell Grants it is less clear how student retention and success have been affected. As early as 1981, a study of the impact of Pell Grants on higher education in Puerto Rico stated that:

> The majority of the students are limited to a cycle of lower division courses, course withdrawals, and eventual dropping out from higher education. This fact negates the apparent increase in higher education training that is suggested by the enrolment increases. In 1978-79, 29.9 per cent of college-aged people were in institutions of higher education, but the data suggests that only 30 per cent of these students graduate, (Duncan, 1981, p.28).

The rate of student success remains a concern. For the cohort of undergraduate students who entered four-year institutions in 1998 there was a remarkable disparity in graduation rate six years later, between public and private institutions. Just under 22 per cent of those in the private sector completed their degrees in that time, while 40.6 percent of students in the public sector did so. Likewise a study of the cohort of students who entered two-year institutions in 2001 indicated that 54 percent of students in the private sector completed their degrees within 150 per cent of the expected completion time, but 66 per cent of those in public institutions (PRCHE, 2004). These data have fuelled a public debate over whether the rapid expansion of the system via Pell Grants has created a 'revolving door' in some higher education segments. While scholarship in this area is relatively new, two main explanations have been advanced for the apparent disparity in success rates between public and private institutions. First, some suggest the disparity is driven in part by relatively high faculty-student ratios in private institutions. Stewart notes

> In terms of student-faculty ratios, Puerto Rican public institutions are close to the US average. Full-time equivalent (FTE) student per FTE faculty for Puerto Rico's public institutions in 2001 averaged 15.5 versus 14.6 for the United States. However, the private institutions in Puerto Rico reported a student-faculty ratio of 26.6 compared with 12.5 for private US institutions (Stewart, 2006, p. 16).

Second, Stewart has noted the level of instructional expenditures in the two sectors. 'For current direct instructional expenditures, Puerto Rico's public universities are spending about 75 per cent as much as their US counterparts per student, while collecting only 29 per cent as much in tuition and fees' (Stewart, 2006, p. 16). Meanwhile Puerto Rico's private institutions spend only 18 per cent as much per student as their US counterparts, and collect 35 per cent in tuition and fees (p.15).

CONCLUSION

The remarkable transformation of the Puerto Rican higher education system after the initiation of Pell grants offers important data for institutional leaders and state and national policy makers around the world as they contemplate a gradual shift in funding allocations from dollars for institutions to dollars for students. After Puerto Rican students received a vast infusion of portable financial aid dollars in 1974, the number of private postsecondary institutions grew rapidly. These institutions were so successful in attracting student enrolments that the balance of public/private enrolments (55/45 in 1974) was reversed in less than a decade and stands now at 35/65. While the number of public institutions and the number of students enrolled in those institutions has grown steadily over the past three decades, private institutions have outpaced the publics, and they have become more concentrated, with 84 percent of private enrolments in eight of the fifty-three private institutions.

New subsidy dollars directed to students can substantially shift the array of institutions students may choose from. In Puerto Rico Pell grants fuelled modest enrolment and revenue growth in existing institutions; and a significant increase in the number of providers. The new institutions, mostly private, have problematic rates of retention and graduation compared to their public counterparts.

In conceptualizing choice-driven education, theorists from Milton Friedman onwards have focused on the prospect of gains in efficiency and effectiveness from greater competition. Attention has only recently been turned to the impact on the production of public goods and the public interest in higher education. Public institutions have been historically essential for low-cost access, social equity, and the production of essential postsecondary products and outcomes that may have low-market demand (Pusser, 2006; Marginson, 2004; Slaughter & Rhoades, 2004). The evidence that rapid increases in subsidies to students have led, in at least one case, to exceptional growth in private non-profit and for-profit institutions is a matter that calls for additional research. Particular attention should be turned to the quality of institutions that have arisen in Puerto Rico in the aftermath of the creation of Pell Grants and the range of majors, curricular offerings and the like in those institutions. As early as 1981 R.J. Duncan expressed the following concern over the evolution of the Puerto Rican postsecondary system:

> Federal student financial assistance has been a primary policy force during the last decade, and most local decisions on higher education have been derivative from the premise of that financial assistance. This Federal assistance has been designed to go directly to the student, rather than as direct institutional grants. So, the pressure on the institutions of higher education is to admit more students to get the greatest amount of benefit from the Federal aid. The result has been the rapid expansion of student enrolments and subsequent increases in institutional income. In some instances physical facilities and personnel have been stretched beyond educationally acceptable limits to absorb these increases (1981, p. 15).

In the twenty-five years since Duncan's observation the pattern of consumer-driven expansion has continued. At the same time, it has become increasingly clear that without State strategic planning, regulatory guidance and financial support, the Puerto Rican system is not effectively positioned to drive the Commonwealth's research and development aspirations or to build the individual human capital and professional expertise necessary for the twenty first century. Many market advocates around the world hope for a greater degrees of student choice and a reduction of State support to public postsecondary institutions. If current trends continue those wishes may come true. But if the case of the Pell grant-driven system in Puerto Rico is an indicator of the future, this may be wishful thinking.

REFERENCES

Breneman, D. W. (1992). Guaranteed studentl: Great success or dismal failure? United Student Aid Fund, May.

Breneman, D. W. (2003). Declining access: A potential - if slow moving - train wreck. National CROSSTALK, (11)2.

Breneman, D. W. (2005, Entrepreneurship in higher education. New Directions for Higher Education, 129, 3-9.

Burd, S. (2001). Bringing market forces to student loan programs. *The Chronicle of Higher Education*, 47, 2 February, A26.

Burd, S. (2004). Selling out higher-education policy? *The Chronicle of Higher Education*, 50(47), 30 July, A16.

Collis, D. (1999). When industries change revisited: New scenarios for higher education. *Forum for the Future of Higher Education*. New Haven: Yale University.

Departamento del Trabajo y Recursos Humanos (DTRH) (selected years). *Empleo y desempleo*. ELA, Puerto Rico: DTRH.

Dietz, J. L. (1987). *Economic history of Puerto Rico*. New York: Princeton University Press.

Doane, D. & Pusser, B. (2005). Entrepreneurial organization at the academic core: The case of summer sessions. *New Directions for Higher Education*, 129. 43-54.

Duderstadt, J. J. (1999). Can colleges and universities survive in the information age? In R.N. Katz (Ed.) *Dancing with the devil*. San Francisco: Jossey-Bass.

Duncan, R.J. (1981). Federal student assistance policy and higher education in Puerto Rico. *Metas*, 2(1), 15-35.

Fernández, J.; Aponte, E., Díaz, E. y Ramos, L. A. (1999). *Working Group 1*, Puerto Rico Council on Higher Education (PRCHE), 23 March. San Juan: PRCHE.

Geiger, R. L. (2004). *Knowledge and money*. Palo Alto: Stanford University Press.

Iturralde, L. (2005). *Desarrollo de los estudios de posgrado en Puerto Rico y su alcance en la investigación*, December. San Juan: PRCHE.

Kelly, K. F. (2001). *Meeting needs and making profits: The rise of for-profit degree granting institutions*. Denver: Education Commission of the States.

Kirp, D. L. (2003). Shakespeare, Einstein and the bottom line. Cambridge: Harvard University Press.

Kirp, D. L (2004). Access denied. *Crosstalk* 12(1), Winter.

Levin, H. M. & Belfield, C. (2004). *Vouchers and public policy: When ideology trumps evidence*. NCSPE.

Levin, J. S. (2001). *Globalizing the community college*. New York: Palgrave.

Marchese, T. (1998). Not-so distant competitors: How new providers are remaking the postsecondary marketplace. *AAHE Bulletin*, May/June.

Marginson, S. (2004). Competition and markets in higher education: a 'glonacal' analysis. *Policy Futures in Education,* 2(2), 175-204.

Mercer, C. (2005). Federal Pell grants: Implications for increasing the maximum award. *CRS report for Congress*. Congressional Research Service: The Library of Congress, Order Code RL32923. Washington, D.C.

Newman, F. (2001). *Higher education in the digital rapids*. February. Accessed at: www.futuresproject.org.

Puerto Rico Council on Higher Education (PRCHE) (1972). Matrícula y grados conferidos en las instituciones de educación superior de Puerto Rico. San Juan: PRCHE.

Puerto Rico Council on Higher Education PRCHE (selected years). *Matrícula y grados conferidos en las instituciones de educación superior de Puerto Rico*. San Juan: PRCHE.

Puerto Rico Council on Higher Education, PRCHE (1997). *Análisis de los programas de asistencia económica para estudios universitarios*. January. San Juan: PRCHE.

Puerto Rico Council on Higher Education, PRCHE (2000). *Educación superior en Puerto Rico: Hacia una visión de futuro*. Documento Base. San Juan: PRCHE.

Puerto Rico Council on Higher Education, PRCHE (2004). *La educación superior en Puerto Rico*. San Juan: PRCHE and IESALC.

Pusser, B. (2006). Reconsidering higher education and the public good: The role of public spheres. In W. G. Tierney (Ed.), *Governance and the public good*. Albany: State University of New York Press.

Pusser, B., & Doane, D.J. (2001). Public purpose and private enterprise: The contemporary organization of postsecondary education. *Change*, January/February, 18-22.

Pusser, B., Gansneder, B.M., Gallway, N., & Pope, N.. (2005). Entrepreneurial activity in nonprofit institutions, *New Directions for Higher Education*, 129, Spring, 27-42.

Pusser, B., & Turner, S.E. (2004). Nonprofit and for profit governance in higher education. In R. G. Ehrenberg (Ed.), *Governing academia*. Ithaca, NY: Cornell University Press.

Quintero, A. H. e Irizarry, R. L. (2006). *Estudios de casos nacionales: Puerto Rico*. Unpublished monograph.

Ríos, E. (2005). *Los estudios de postgrado en Puerto Rico. Primer informe nacional*. Venezuela: IESALC, UNESCO.

Ruch, R. S. (2001). *Higher education, inc.: The rise of the for-profit university*. Baltimore: The Johns Hopkins University Press.

Schmidt, P. (2001). Texas governor proposes radical shift in college financing. *The Chronicle of Higher Education*, 12 January. Available at: [http://chronicle.com].

Slaughter, S., & Rhoades, G. (2004). *Academic capitalism and the new economy*. Baltimore: Johns Hopkins Press.

Stewart, J. R. (2006). Benchmarking Puerto Rico's education systems. *Boletín CESPR*, 7(1), June.

Suspitsin, D. (2002) Pell grants. In: J.J.F. Forest and K. Kinser (Eds.), *Higher education in the United States, an encyclopedia, Volume 2*. Santa Barbara, CA: ABC Clio Press.

Texas State Senate Special Commission on Twentyfirst Century Colleges and Universities (2001). *Moving every Texan forward*. Austin.

Sperling, J.G. (2000). *Rebel with a cause: The entrepreneur who created the University of Phoenix and the for-profit revolution in higher education*. New York: John Wiley.

Tooley, J. (2001). *Reclaiming education*. New York: Cassell.

United States Department of Education Budget Service (2007) *Highlights of the FY 2007 education budget*. Department of Education: Washington, D.C.

United States Department of Education (2007) *Overview, Fiscal Years 2001-2007, State tables for the United States*. Department of Education: Washington, D.C.

Vélez C., W. (2002). *El financiamiento de la educación superior en Puerto Rico*. San Juan: Ediciones Educación Superior.

Vélez C., W. (2006). La reestrcturación neoliberal de la UPR y el aumento en los costos de matrícula y cuotas. *APUESTA*, 1, Febrero, 25-32.

Weiler, H. N. (2001). *States and markets: Competing paradigms for the reform of higher education in Europe*. New York: National Centre for the Study of Privatization in Education, Teachers College, Columbia University.

Winston, G. (1999). For-profit education: Godzilla or Chicken Little? *Change*, Jan/Feb.

Gloria Dávila-Casasnovas
Puerto Rico Council on Higher Education.
Puerto Rico

Brian Pusser
Centre for the Study of Higher Education, Curry School of Education
University of Virginia, Charlottesville, United States of America.

Waldemiro Vélez-Cardona
Department of Social Sciences, University of Puerto Rico,
Recinto de Rio Piedras, Puerto Rico

GARY RHOADES, ALMA MALDONADO-MALDONADO, IMANOL ORDORIKA & MARTÍN VELAZQUEZ

8. IMAGINING ALTERNATIVAS TO GLOBAL, CORPORATE, NEW ECONOMY ACADEMIC CAPITALISM

INTRODUCTION

In a post-Soviet world, there seems little alternative to a market-based economy. From one country to the next, the prevailing political ideology is neo-liberalism, promoting a marketizing of public sector enterprises. In higher education this means there is little alternative to what Slaughter & Leslie (1997) and Slaughter & Rhoades (2004) have called academic capitalism. The intersection of the new economy, global corporations, and universities is most studied in the advanced Western world (Clark, 1998; Currie & Newson, 1998; Marginson & Considine, 2000; Bok, 2003). At the turn of the twentieth century the U.S. model of entrepreneurial universities and of engagement with the corporate world and the marketplace has become a dominant policy model of higher education, just as the German Humboldtian model was at the turn of the last century. And in both cases, the model that is promoted, and eventually adapted, is often in many ways unlike the original model. This recalls Amy Chua's *World on fire* (2003) where she argues that Western/ U.S. models of the market economy and of democracy are adapted in developing countries in ways diverging significantly from their origins, particularly the absence of institutional infrastructures surrounding the models.

The U.S. model continues to be aggressively promoted by international entities such as the World Bank as the only feasible and realistic policy choice and route to growth for developing countries. And just as clearly, oftentimes the move to the market in the higher education systems of other countries in some regards proceeds much further and to a greater, unrestricted extreme than it has in the U.S. Yet we believe there are realistic and realizable alternatives to the prevailing American model of academic capitalist universities. In this chapter we draw particularly on examples from Latin America to imagine alternativas.

We begin by briefly setting out the conditions and patterns of academic capitalism and the new economy in U.S. higher education. In the process, we identify a few problems with that model, related to what constituencies and communities of entrepreneurial universities prioritize. Subsequently, we detail a conceptual model for considering the international reach and the national and local patterns of academic capitalism. Then, we offer an example of a distinctive Mexican entrepreneurial alternative, with a different balance of structures and

Simon Marginson (ed.), Prospects of Higher Education, 171–182.

commitments, in a leading private university. We also find alternatives embedded in European traditions of Latin American public universities, which offer the possibility of a forward looking conception of a universidad latinoamericana that is different in important ways from the prevailing academic capitalist model. Such an alternativa would address the varied cultural, political, social and economic needs of the local communities and nations in which the institutions are positioned, while creating a space for a modelo distintivo global of what a university can be.

Academic Capitalism

Universities have long engaged in market-like behaviour. However, recent decades have witnessed the emergence of 'academic capitalism in the new economy', a regime of policies and practices that involve colleges and universities engaging in market and market-like behaviours (Slaughter & Rhoades, 2004). Those practices extend to higher education's central research, educational, and service functions. In the U.S., colleges and universities are not simply partnering with business, they are developing, marketing and selling products commercially in the private sector marketplace. Essentially, they are looking to generate profits, whether through patents and technology transfer, through the technologically mediated delivery of education and educational materials, or through non-academic consumption items. The move in this direction has involved a revision and creation of policies to make these activities possible. The relationship between states, their higher education institutions and private sector organizations is changing, as the boundaries between the for-profit and not-for-profit sectors are being blurred. Within higher education this has meant foregrounding market and potential revenue generation in policy negotiation and in strategic and academic decision making.

The changes embedded in academic capitalism and in the new economy are not simply about revenue generation. They are also about the new, so-called information based economy. That has meant the development of new networks with high tech businesses and industries.

At the same time as these developments in U.S. higher education have taken place, we have witnessed the ascendance of neo-liberal policies nationally and internationally. Such policies shift government investment in higher education. They emphasize college and universities' economic role and cost efficiency. They involve institutions aiming to generate more of their own monies, to connect more closely to private corporations, and to be more efficient in and accountable for their use of public monies. Over time, then, conceptions of higher education as a producer of public goods and a cultural project have become more marginal, in the face of a conception of higher education as a private good.

What this has meant in the U.S. is a shift in institutions' orientations in important ways, not only to employers but also to 'customers.' The focus on revenue generation has meant an increased emphasis on connections with large corporations, as distinct from small and medium sized corporations. The new economy focus has meant as well a shift in the types of business and sectors of industry with which higher education intersects. It has also meant a shift in the target student populations, towards students with a greater ability and willingness to pay more to gain access to higher education (Slaughter & Rhoades, 2004). Academic capitalism in the new economy is not neutral about who it serves. It focuses attention on the highest bidders, among employers and students. Perhaps the most dramatic example of this is among community colleges, which are increasingly intersecting with international student markets; the colleges can charge international students higher tuition. The 'community' orientation in terms of student access is shifting not only with the pursuit of international students but also with the growth of distance education. Again, one of the reasons for this is that colleges can charge distance students higher fees. There is also a hope that high tech delivery costs will reduce labour costs. In short, the pattern of academic capitalism in the new economy encourages institutions to incorporate more of their activities into the global economy and in relative terms to disinvest in activities, programs, and students that are centred in the immediate locale and region.

Influencias y Mecanismos Globales de Capitalismo Académico

One of the principal perspectives for conceptualizing and analyzing globalization is to consider it as the 'natural continuation of capitalism', as 'another stage of its development' (Waters, 1995). According to this perspective, globalization relates directly and naturally to neo-liberal policies and practices. However, we do not see the connection as one that naturally unfolds in a structured sequence of stages of capitalism. Rather, we focus on the international agencies and human agency that are involved in mediating and translating influences between international, national, and local social systems (Marginson & Rhoades, 2002). In particular, in this section, we focus on the role of international organizations in shaping higher education policy and practice. These organizations provide a useful way of examining key social actors who shape the spread of neo-liberal higher education policies. As Scott (1998, p.122) notes, 'not all universities are particularly international, but all are subject to the same processes of globalization - partly as objects, victims even, of these processes, but partly as subjects, or key agents, of globalization'. In this analytical spirit we examine mechanisms and patterns of academic capitalism in Latin America.

Latin America is characterized by great variety in higher education systems and institutions. For all the variety, however, it is necessary to consider a commonality, which is the inequalities within the systems, and particularly the inequalities between them and the higher education system of the U.S. The U.S. model, so evident in the policies of international organizations, has a powerful influence on Latin American institutions and systems of higher education. In some regards Latin American higher education adapts key dimensions of this model and in others actively resist it. For our purposes, we consider the three 'megasystems' in the region: Argentina, Brazil, and Mexico (Garcia Guadilla, 2000), significant because of their size. We also consider Chile because it was the first country in the region to establish neo-liberal reforms, and can be seen as the laboratory of the World Bank's efforts to shape Latin American higher education systems.

Since World War 2, international organizations have had a significant presence in the region. The most global of these have been the World Bank, the United Nations Educational, Scientific, and Cultural Organization (UNESCO), and the Organization for Economic Cooperation and Development (OECD). Significant regional organizations affecting higher education have included the Inter American Development Bank (IDB), the International Institute for Higher Education in Latin America and the Caribbean (IIHELC-UNESCO), the Economic Commission for Latin America and the Caribbean (ECLAC), the Organization of American States (OAS), and the Iberia-American States Organization for Education, Science, and Culture (IASO) (Maldonado 2003). These organizations have published documents that have had a major impact in the region (CEPAL, 1990; Task Force on Higher Education and Society, 2000; Winkler, 1994; World Bank, 2002).

Although reforms have taken different form from one country to the next, there are important common patterns. Throughout the region the reforms have promoted three interrelated neo-liberal policies: reduced public financing, evaluation and quality assurance, and increased tuition and privatization as a means for providing access. Each reflects the influence of the U.S. model of academic capitalism.

During the 1980s Latin America had the lowest public financing of higher education of all world regions (Brunner et al., 1994). Moreover, governments are increasingly seeking to raise tuition in the public sector, shifting the costs to the consumer. The trend of low public finance continued through the 1990s and became used by governments to impose stricter controls on resource allocation to universities, such as by linking financing to evaluation. National quality control bodies have been created everywhere, reflecting the global pattern of an 'evaluative state' (Neave, 1998) that effects change through the mechanism of intermediate evaluative bodies. The models of quality assurance and accreditation the result reflect some features of the U.S. model (Rhoades & Sporn, 2002), but taken to a more restrictive and more national level than in the U.S. Thus, in Argentina, a national body was established in 1995, Comisión Nacional de Evaluación y Acreditación Universitaria. In Brazil, three bodies have been established: the National Council of Higher Education Evaluation (1993), the National Institute of Studies and Educational Research (1995), and the National Evaluation System (2002). In Chile, four councils were established: Consejo Superior de Educación (1990), Comisión Nacional de Acreditación Pregrado (undergraduate) (1998), Comisión Nacional de Acreditación Posgrado (1998), and the Sistema Nacional del Aseguramiento de la Calidad (quality assurance) de la Educación Superior (2002). In Mexico, four bodies have been created since 1989: Comisión Nacional de Evaluación (1989), Comités Interinstitucianales para la Evaluación de la Educación Superior (1991), Centro Nacional de Evaluación (1994), and the Instituto Nacional para la Evaluación de la Educación (2002).

For all the emphasis on quality, there is an interesting contradiction that emerges with respect to private higher education in Latin America. As governments have restricted the support and growth of public institutions, the demand for greater access has been accommodated by the expansion of private institutions and by new, non-university institutions. In Mexico from the mid 1980s to 1999 only three public universities were created, as compared to 45 technological universities and 35 private universities. The private institutions generally charge relatively high tuition. They are not subject to the same quality assurance mechanisms as public institutions, leading to substantial questioning of their quality. The latter point is particularly true in Brazil. Both the privileging of private institutions in public policy, and the promotion of higher tuition in the public sector, again reflect features of higher education in the U.S.; although there, privates are privileged through financial aid policies, and non-university public institutions - community colleges - are the principal point of access for low income (and Latino) students.

To what extent are the reforms shaped by international organizations? Analysis of their documents reveals a clear neo-liberal agenda (Maldonado forthcoming). International organizations call for restriction of public financing of higher education, for more diverse and more private higher education, for quality assurance mechanisms that link resource allocations to evaluation, and for shifting the costs of higher education from the state, as an investment in the public good; to the consumer, as an investment that is a private good and benefits the individual. The mechanisms of influence are both coercive (for example the World Bank's structural adjustment policies: see Kempner & Jurema, 2002) and normative. The latter processes involve the dissemination of publications that norm the Latin American national systems on an international standard and in relation to each other, the creation of policy networks and the empowering through research and contracts of particular groups that shape national higher education policy. These are important mechanisms for influencing public policy in developing countries.

The connection to the academic capitalism model in the U.S. is plain. Despite the World Bank's claim of neutrality, social scientists and practitioners have long recognized its direct link to the U.S. 'In the first fifty years, the Bank did not begin to escape the charge that the United States had not only the heaviest weight but, compared with the legalities, a disproportionate one in the governance of the institutions' (Kapur et al., 1997, p. 3). Perhaps more importantly, the organizations promote a neo-liberal, market model that is premised on the success of the U.S. higher education system and the ability to transfer it directly to other countries.

Yet as noted, what is transferred and applied often bears little resemblance to the U.S. model in practice. For example, private institutions in the U.S. are subject to accreditation and quality assurance mechanisms, but for the most part in the U.S. these mechanisms are not directly linked to resource allocation. Access to higher education is primarily accomplished through low tuition public institutions of higher education, and the federal government provides enormous subsidies to facilitate the shifting of higher education costs to the consumer. None of these conditions apply in Latin American systems of higher education.

Finally, whatever the influence of the international organizations and the U.S. model, the mechanisms through which that influence is translated into actual policies and practices in national and local settings depends on national and local 'experts'. Such individuals and groups, which can be conceptualized as 'epistemic communities', networks of professionals in policy domains (Haas, 1992), play an important role in shaping policy. It is important to consider their position in regard to academic capitalism, U.S. style (Maldonado, forthcoming). For they are a key to understanding the distinctive adaptations of academic capitalism in Latin America.

Una Alternativa Distintiva de Universidad Privada Mexicana

One leading private universities in Mexico shows how the character of academic capitalism can vary significantly from one context to the next. Entrepreneurial activity can go in some very different directions from what the neo-liberal model of higher education that is promoted by the international agencies would suggest.

The Instituto Tecnológico y de Estudios Superiores de Monterrey (ITESM), commonly known as Monterrey Tec, is the largest private university in Mexico. It is a university system with 33 campuses located in different cities and states. It offers undergraduate, masters, and doctoral degrees. It has about 100,000 students. Founded in 1943 by a group of businessmen concerned about a perceived lack of highly skilled professionals in Mexico, Monterrey Tec was explicitly modelled on the Massachusetts Institute of Technology, which the Tec's creators had visited. This private university is one of only three Mexican universities accredited by the Southern Association of Colleges and Schools. It also has the recognition of the U.S. Department of Education and is a member of the College Board. Here one can already see some of the international influences discussed in the previous section of the paper, although the particular mechanisms of influence are different.

Despite the number and strength of international influences, or more accurately in this case of U.S. influences, on Monterrey Tec, it nevertheless presents a distinctive model of private higher education. ITESM is a private, non-profit university that operates in a country in which the public higher education sector is dominant and comprehensive, whereas the private sector is peripheral and more focused in orientation, though growing rapidly (Geiger, 1988; Levy, 1986). Monterrey Tec has focused its curriculum on professional degrees targeted to private sector employment. However, in contrast to private universities in the U.S. such as MIT, it is focused on undergraduate education, and is characterized by relatively limited involvement in doctoral education and research, including applied research, particularly outside the Monterrey campus (it is interesting in this regard that it is not modelled more explicitly on selective liberal arts colleges in the U.S.). Still, it has adopted an orientation toward the private labour market that is far more explicit in various regards than is the case for public Mexican universities.

Another interesting contrast is found in the funding structure and configuration of revenue streams on which Monterrey Tec relies. Private higher education institutions in Mexico do not receive any direct financial support from public agencies. That makes for a substantial contrast with elite private universities in the U.S., which receive a large amount of public support directly in the form of research grants, and a great deal of indirect support through publicly underwritten financial aid. For Mexican private universities generally, and for ITESM, the principal revenue stream is from student tuition. The only financial aid comes from the institutions themselves: in the case of Monterrey Tec, 33 per cent of its students receive such aid. There are preferential tax schemes in Mexico so that tuition for undergraduate programs is tax exempt. Tuition revenues are not taxed either. Private philanthropic donations are also tax deductible, and Monterrey Tec collects capital through fund raising campaigns. But these fund-raising campaigns pale in comparison to those of large public and private universities in the U.S., with most of the monies coming from members of ITESM's Board. In fact, the second largest source of revenue for Monterrey Tec is not private gifts, but is the Sorteo Tec, which is nation-wide authorized lottery, for which companies donate prizes.

Two other distinctive features of Monterrey Tec, in comparison to its private sector counterparts in the U.S., lie in its commitment to and investment in global exchange and sustainable local community development. As with most selective private colleges and universities in the U.S., Monterrey Tec has large study abroad and exchange programs. Partly because it wants its students to develop proficiency in English to enable them to gain employment in the global economy, ITESM encourages its students to spend at least one academic term abroad. More than 7,000 students participated in such a program in 2003. ITESM has exchange programs with over 300 universities throughout the world, and has eight liaison offices in the U.S., Canada, and Europe. These activities are not undertaken to generate revenue. No service fees are charged to students who study abroad.

Likewise social service is part of undergraduate education in Mexico and Monterrey Tec has a targeted program in this area. Since 1945, by federal law undergraduate students in Mexico must complete at least 480 hours of social service, including activities related to students' professional major or community service. The activities can be carried out at the student's institution, or in the public or private sector. Higher education institutions are responsible for their own social service program. The ideal is one in which the institution links its students with strategies to solve problems affecting marginal communities (Bertín, 2000; Mungaray & Oceguedo, 1999). Monterrey Tec's program requires that a minimum of 240 hours – in most campuses it is 300 hours out of the total of 480 – are devoted to social community service. Because the institution focuses primarily on social service rather than internship during the 480 hours, most students' practical work in their intending profession is built into their formal coursework program.

ITESM undergraduates are encouraged to participate in activities promoting the development of marginal communities, focusing on education, entrepreneurial skills, environment and ecology, health and hygiene, housing and technical training (ITESM, 2001). Students take an induction workshop to acquire understanding of community service. Ernesto Benavides, Director of the Social Community Service program at ITESM points out (personal interview) that above all, in these programs Monterrey Tec uses a model of self-management. Rather than paternalistically providing temporary aid and relief it fosters sustainable small business enterprises in poor communities, not unlike the micro-lending programs of some nongovernmental organizations. In 1997, ITESM began to focus the social community service activities of each campus onto a small number of communities in order to achieve continuous, long-term development. It is now also linking these programs more systematically with faculty members and academic programs.

The activities that fall within community service programs are wide ranging. They entailed in the last year over three million hours of student time. A few examples will indicate the sort of work that is done. One program involved the development of a denim clothing manufacturing business run by women in the community. Another involved the reconstruction of a medical clinic. Yet another project entailed the construction of housing in poor communities, with the families themselves participating in the construction process and learning various skills, not unlike the model of Habitat for Humanity. In a further case, students organized a literacy and math program for rural cheese producers, to enhance their business skills. Still another project resulted in the creation of a 'Fundación la Paz Comienza con los Niños' that teaches courses in arts, sciences, and physical education to 750 children from foster homes. Each of these activities targets poor local communities with the aim of establishing services and skills that will endure well beyond the life of the students' and institutions' activities. Many might question how successful and enduring these efforts are; and as the organization of targeted activities is relatively recent, in most cases it is probably too early to tell. What is important in relation to this chapter is that this elite private university provides an example of how a university can invest in, enhance, and work to effect sustainable change and improvement of the quality of life in local communities.

Developing a Modelo Distintivo de Universidad Latinoamericana

Historically Latin American higher education acquired meaning and purpose within a long tradition established after the Universidad de Córdoba (Argentina) reform in 1918. The most salient features of this tradition have been university autonomy, shared governance, tuition-free education and strong commitment to the university's social and political responsibilities in national development, sovereignty and the defence of democracy. Notwithstanding at least three waves of university transformations, during the 1930s, 1940s, and late 1960s/early 1970s, Latin American universities maintained their unique identity in the face of strong influences from the university traditions of the United States. Nearly forty years ago Darcy Ribeiro (1969) pointed to attempts to establish an 'Americanization' project in Latin American higher education, citing debates about humanism or practical orientations, elitism or massification, and science versus professionalism.

However, in the face of contemporary trends in higher education reform Latin American universities have lost much of their traditional identity. Universities have been forced by structural conditions, including reduced public funding and privatization, to adopt policies and strategies deemed successful by international experts and agencies in the establishment of entrepreneurial institutions. In most Latin American countries the identities, meaning, and purposes of higher education institutions are in the process of being lost in attempts to imitate the policies and practices of internationally acclaimed research universities. Throughout Latin America, particularly in Mexico, universities have been involved in marketization and privatization, diversifying their sources of funding and competing for financial resources (Mollis 2003). Private investment in higher education has increased and for-profit providers have emerged. Public institutions themselves are becoming service providers and are attempting to increase university-business connections. As outlined earlier, most countries have established complex evaluation systems, accountability measures, and accreditation and certification procedures for institutions, academic programs, and academics (Ordorika, 2004). These systems are geared towards competition for public funds, and merit pay income for faculty (Canales Sánchez, 2001; Diaz Barriga & Pachec,o 1997; Ordorika, 2004). Despite limited private sector markets for the production of knowledge and professionals, higher education systems and institutions have adopted a market discourse and attempted to assimilate market practices into academic systems (Ordorika, 2002).

To some extent, changes in policies and practices in Latin America have been 'structurally adjusted' as part of World Bank and International Monetary Fund (IMF) packages to address the fiscal and debt crises of countries (Carnoy & Samoff 1990). At the same time, these changes have been quite politicized and in many cases have been strongly resisted. During the last twenty years, intense political confrontations have been a significant element of university life. In the face of tuition increases, privatization, and the loss of meaning and identity, universities have been the site and object of political conflicts, student movements, and faculty strikes. Most of these confrontations have been focused on resisting what are commonly labelled as neo-liberal policies towards higher education.

Yet often resistance has defended a mystification or romanticization of Latin American universities. Critical stances and understanding of the limitations of higher education institutions and policies in Latin America have been displaced by nostalgic attempts to 'restore' university traditions and practices that are part of a myth. Romanticization of the history leaves little scope for going beyond resistance to create a new, indigenous identity for Latin American higher education.

At the same time, attempts to mimic entrepreneurial Western (and particularly American) universities in the face of scarce resources, extremely constrained markets, economic subordination and technological dependence, produce grotesque outcomes. 'Underdeveloped universities' try to resemble and compete with highly developed core universities in knowledge production, technology development and transfer, and cutting edge research. In aspiring to emulate the core universities, universities in Latin America increasingly turn towards serving international and corporate markets and away from serving public sector markets and the public good. But there is a need to explore the nature and role of colleges and universities in peripheral, developing and underdeveloped regions in Latin America.

The creation of Latin American university identities, projects, and practices should be rooted in an historical understanding of, and commitment to, redressing the economic, political, and social realities of these countries. Latin American universities have long played a central national role in their societies. In a time of globalization, powerful international influences are pushing them to become academic capitalists, primarily intersecting with global markets at the expense of local and national needs, including economic ones. The challenge is for Latin American universities to use their historical role as a touchstone, on the basis of which they can identify critical issues that represent national projects to which higher education institutions can commit and in which they can invest. Alternatives to academic capitalism lie in alternatives to current patterns of neo-liberal development which involve sublimation to international business.

We identify three such alternatives as significant projects. One is focused on democratization, not only of access to higher education but of governance and politics in society. How can higher education be linked more to the promotion of increased access as a public good, and support it more as a public investment? How to feature more prominently and support more systematically the democratic role of universities in Latin America, not only as social critics or as political insiders, but as institutions that contribute to strengthening institutional foundations in local, state, and national communities of democratic processes, debates, and practices?

A second project is that of independent development, not just economic but social. In what ways can Latin American universities contribute to the sustainable economic vitality of local and regional communities and of the nation as a whole? How can these universities play a role in expanding infrastructures that support public health, literacy, and education?

Finally, a third project is that of sovereignty and cultural distinctiveness. In pursuing their distinctive identities, Latin American universities should consider how they might contribute to building new and sovereign cultural identities in their countries, in ways that draw on the past but also look to the future. And they should commit to the development and fostering of new economy commerce that expresses, advances, and contributes to strengthening that identity, whether in the media and fine arts or in the humanities. Several years ago, one of the co-authors heard an English professor at a conference in the UK argue that what was most marketable in Britain in the global economy was not technology, but high culture. Global markets are more likely to find Jane Austen, Shakespeare, the BBC, and other forms of English culture more attractive than British technology. Without arguing that either the UK or Latin American universities ought not to play the science and engineering game, at least in global markets, there is much room for a greater emphasis currently on the sovereign identities and cultural contributions of Latin American. For example, as with the UK, what indigenous high technology has been as successful globally as the literary contribution of magical realism?

We suggest that Latin American universities play to their strengths and to their historical national roles in developing a modelo distintivo de universidad latinoamericana. The most feasible and successful future for these institutions lies not in emulating academic capitalism, U.S. style. It lies in drawing on their own distinctive strengths in addressing the challenging national projects of the future.

REFERENCES

Bertin, G.I. (2000). *El servicio social en México. Nacional de Universidades e Instituciones de Educación Superior*. Access at: http://www.anuies.mx/principal/servicios/publicaciones/libros/lib50/43.htm]

Bok, D. (2003). *Universities and the marketplace: The commercialization of higher education*. Princeton: New Jersey: Princeton University Press.

Brunner, J.J., Balán, J., Courard, C., & Cox, C. (1994). *Educación superior en América Latina: Una agenda de problemas, politicas y debates en Unbral del año 2000*. Santiago, Chile: FLACSO.

Canales Sánchez, A. (2001). *La experiencia institucional con los programas de estimulo: La UNAM en el periodo 1990-1996*. México, DF: DIE, CIN VESTAV.

Carnoy, M., & Samoff, J. (1990). *Education and social transition in the third world*. Princeton, New Jersey: Princeton University Press.

CEPAL. (1990). *Transformación productiva con equidad*. Santiago, Chile: CEPAL.

Chua, A. (2003). *World on fire: How exporting free market democracy breeds ethnic hatred and global instability*. New York: Doubleday.

Clark, B. R. (1998). *Creating entrepreneurial universities: Organizational pathways of transformation*. Oxford: IAU Press & Pergamon.

Currie, J., & Newson, J. A. (Eds.) (1998). *Universities and globalization: Critical perspectives*. Thousand Oaks, California: Sage.

Diaz Barriga, A., & Pacheco, T. (1997). *Universitarios, institucionalización académica y evaluación*. pp.210 in Pensamiento Universitario, 86, editado por. Coyoacán Universidad Nacional Autónoma de México Coordinación de Humanidades Centro de Estudios Sobre la Universidad.

Garcia Guadilla, C. (2000). Latin America: Higher education research in a transformation context. In U. Teichler & J. Sadlak (Eds.) *Higher education research: Its relationship to policy and practice*. Oxford: IAU Press, Pergamon/UNESCO.

Geiger, R.L. (1988). Public and private sectors in higher education: A comparison of international patterns. *Higher Education*, 17, 699-711.

Haas, P.M. (1992). Introduction: Epistemic communities and international policy coordination. *International Organization* 46 (1), 1-35.

ITESM (1996). *Misión del sistema tecnológico y de estudios superiores de Monterrey*. 6 de septiembre.

ITESM (2001). *Reglamento general del servicio social comunitario y profesional para los alumnos de sistema tecnológico de Monterrey*.

Kapur, D., Lewis, P., & Webb, R. (Eds.) (1997). *The World Bank: Its first half century*. Washington, D.C.: Brookings Institution Press.

Kempner, K., & Jurema, A. L. (2002). The global politics of education: Brazil and the World Bank. *Higher Education*, 43(3), 331-54.

Levy, D. C. (1986). *Higher education and the state in Latin America: Private challenges to public dominance*. Chicago: University of Chicago Press.

Maldonado-Maldonado, A. (2003). Investigación sobre organizmos internacionales a partir de 1990 en México. In P. Ducoing (Ed.) *La investigación educativa en México. "Sujetos, actores y procesos de formación", formación para la investigación*. Los Académicos en México, Actores y organizaciones. México: COMIE-SEP-CESU.

Maldonado-Maldonado, A. (forthcoming). *The influence of international organizations in the field of higher education in Latin America: The analysis of one epistemic community in Mexico*. Doctoral dissertation, Boston College.

Marginson, S, & Considine, M. (2000). *The enterprise university: Power, governance, and reinvention in Australia*. Cambridge: Cambridge University Press.

Marginson, Simon, & Rhoades, G. (2002). Beyond national states, markets, and systems of higher education: A glonacal agency heuristic. *Higher Education*, 43(3), 281-309.

Mollis, M. (Ed.) (2003). *Las universidades en América Latina: Reformadas o alteradas*. Buenos Aires, Argentina: CLACSO.

Mungaray, A. & Ocegueda, J.M. (1999). *El servicio social y la educación superior frente a la pobreza extrema en México*. Asociación Nacional de Universidades e Instituciones de Educación Superior.

Neave, Gy. (1998). The evaluative state reconsidered. *European Journal of Education*, 33(3), 265-84.

Ordorika, I. (2002). Mercados y educación superior. *Perfiles Educativos* 24(95), 98-103.

Ordorika, I. (2004). *El mercado en la academia. En la academia en jaque: perspectiva politica sobre la evaluación en México*. In: Ordorika, I. (Ed). México, DF: CRIM-UNAM/Miguel Angel Porrua.

Rhoades, Gary, & Sporn, Barbara (2002). Quality assurance in Europe and the U.S.: Professional and political economic framing of higher education policy. *Higher Education* 43(3), 355-90.
Ribeiro, D. (1969). *A universidade necessaaria.* Rio de Janeiro: Paz e Terra.
Scott, P. (1998). *The globalization of higher education.* Buckingham: Open University Press.
Slaughter, S., & Leslie, L. L. (1997). *Academic capitalism: Politics, policies, and the entrepreneurial university.* Baltimore: The Johns Hopkins University Press.
Slaughter, S., & Rhoades, G. (2004). *Academic capitalism and the new economy.* Baltimore: The Johns Hopkins University Press.
Task Force on Higher Education and Society (2000). *Higher education in developing countries: Peril and promise.* Washington, D.C.: World Bank.
Waters, M. (1995). *Globalization.* London: Routledge.
Winkler, R.D. (1994). *La educación superior en América Latina. Cuestiones sobre eficiencia y equidad.* Washington, D.C.: Banco Mundial.
World Bank. (2002). *Constructing knowledge societies: New challenges for tertiary education.* Washington, D.C.: World Bank.

Gary Rhoades, Alma Maldonado-Maldonado & Martín Velazquez
Center for the Study of Higher Education
University of Arizona
Tucson, Arizona

Imanol Ordorika
Universidad Nacional Autonoma de Mexico
Mexico City, Mexico

PART III

PERSPECTIVES OF NATIONS AND REGIONS

'If a person can have only one identity, then the choice between the national and the global becomes an "all or nothing" contest. And so does the contest between any global sense of belonging we may have and the local loyalties that may also move us. But to see the problem in these stark and exclusive terms reflects a profound misunderstanding of the nature of human identity, in particular its inescapable plurality'.

- Amartya Sen, *Identity and violence: The illusion of destiny*, Allen Lane, London, 2006, p. 182

JUSSI VÄLIMAA & DAVID M. HOFFMAN

9. THE FUTURE OF FINNISH HIGHER EDUCATION CHALLENGED BY GLOBAL COMPETITIVE HORIZONS

INTRODUCTION

In this chapter we reflect on significant processes of change in Finnish higher education with an eye to the future. In order to speculate on the future of Finnish higher education, we intend to illuminate and understand processes that have continuity from the past to present-day, thus having important implications for the future. This point of departure utilizes one of the basic questions asked by historians: What are the continuities and discontinuities in societies?

In this study we formulate this historical starting point into three research questions: What are the most significant traditions of Finnish higher education? What are the aims of current reforms and changes in Finnish higher education system? How do the present-day changes and reforms interplay with significant traditions in a way which gives us insight about our future?

We address these questions from three perspectives. First, we will analyse the traditions that form the context in which present day reforms and changes manifest and may ultimately challenge. Secondly, we will analyse current debates in Finnish higher education related to globalization because economic and cultural globalization inherently challenge Finnish higher education as a national system of higher education. The globalization of higher education is not a mystical or an abstract process. It is presently manifesting in concrete processes of change in and through the Bologna Process taking place in European higher education. Therefore it is necessary to take a look at the Bologna Process and the changes related to it in the Finnish context. Finally, we will look at Finnish higher education in terms of the competitive horizons illuminated by the Bologna Process in relation to both recent changes and with respect to the traditions of Finnish higher education.

TRADITIONS AND HISTORICAL LAYERS

Finnish higher education has its roots in the western tradition of universities. This tradition began in the Middle Ages when a number of Finns were sent to European universities - most often to Paris - returning home to work as bishops or other high clerical officials in Catholic Church in the so called 'Eastern provinces' of the Swedish Kingdom. In order to understand the nature of Finnish higher education it is essential to understand that the Finnish nation state was born out of debates

Simon Marginson (ed.), Prospects of Higher Education, 185–200.

within higher education. The cultural, political and ultimately national identity of Finland emerged from higher education, not the other way around (Välimaa, 2001a). This historical relationship between higher education and society remains in place - in some sense - to this day. Distinct aspects and episodes of this interdependence have left visible features on the functioning of higher education institutions even though it is not always easy to see their origin. We will use the idea of historical layers deposited over time, one upon another, like archaeological layers in order to reveal the features which shaped the formation of Finnish higher education. Earlier historical layers exert continuous influence on modern practices not only because traditions consisting of ideas, norms and practices have provided starting points for new developments but also because they have given form and shape to more recent layers. In academia, traditions are crucially important. All new and emerging knowledge and intellectual activities are related to what has been known and understood before, whether we are aware of those origins or not.

Our understanding of historical layers aims to combine linear passing of time with the Hegelian philosophical tradition of history, the idea that history advances through struggles following the logic of thesis, antithesis and synthesis which results in qualitative change(s). New historical layers are not necessarily created through revolutionary developments; and even when this might be the case, it does not mean that everything is changed with one stroke. Therefore, it is sometimes difficult to see where previous historical layer ends and where a new one begins, just as is the case with archaeological layers. The historical layer as an intellectual device aims to focus attention on continuities with the present-day and times past. With the help of historical layers it is also possible to acknowledge that the past has always been full of details as well as competing traditions and practices. Here the aim is to analyse practices which have historical continuity in the present.

The Historical Layers of Finnish Higher Education

The basic historical layer underpinning Finnish higher education was deposited in the Middle Ages when the Catholic Church linked Finland culturally with continental Europe. This was also one of the factors, together with the Finnish region's incorporation into the Kingdom of Sweden under the name of the Eastern Provinces, that incorporated Finland into the Western European cultural sphere. The expanding Swedish Kingdom required improved training of priests to defend the right (version of the) protestant religion and civil servants for the Swedish King. The Turku cathedral school was enlarged and transformed into the University of Turku in 1640. This formed the second historical layer of Finnish higher education. It also marked the beginning of the foundation of a national (system of) higher education. This social function of higher education as an educator of civil servants continues to be important to this day. Two thirds of university and over half polytechnic graduates become public-sector employees.

The third historical layer making up present-day Finnish higher education accumulated during the time when Finland was an autonomous Grand Duchy of the Russian Empire from 1809 to 1917. This was a historical period when Finland

and the Finnish nation developed culturally, politically and economically. It has been said that Finland developed from a geographical concept into a political concept during that period. The only university in Finland, the Imperial Alexander University (the former University of Turku and later known as the University of Helsinki), played a particularly prominent role in the cultural, political and economic processes that created a basis for the independent nation state which was established during the First World War. Among the essential starting points for the development of a Finnish conception of a university was the idea that a university is a national cultural institution. This view was rooted in the Humboldtian ideal of university which focused attention to moral development of students during their studies and to the institutional autonomy of the university and the academic freedom of its professors. These understandings of the idea of university remain central to today's Finnish universities. In addition, two-year Teacher Training Seminars were established from 1860s onwards. These higher education institutions opened an educational channel for the sons and daughters of peasant and workers during the nineteenth century. School teachers became an important group involved in the education of the emerging nation. The Imperial Alexander University and higher education were integral both culturally and politically to the development of the nation and the nation state (Välimaa, 2001a).

The fourth historical layer underlying Finnish higher education was laid down between the two world wars. At this time Finnish higher education was still an elite system and the number of higher education students remained small (Nevala, 1999). University professors enjoyed a high social status, many of them serving as ministers in the governments during the 1920s and the 1930s. The high social prestige of universities and university degrees is still a social reality in Finland at the beginning of the twenty-first century. In recent national study universities and other higher education institutions scored top marks in comparison with other institutions of Finnish society. As many as 70 per cent of the respondents expressed great confidence, and only 9 per cent expressed little confidence, in higher education institutions (Finnish Science Barometer, 2004).

The fifth historical layer forming Finnish higher education occurred through the expansion of higher education in the direction of a mass higher education system. This development began in the late 1950s with the establishment of the University of Oulu in Northern Finland, followed by the establishments of universities all over Finland by the end of the 1980s. Measured by student numbers, higher education in Finland became a mass system during the 1970s, about two decades earlier than in other Nordic countries. In 2003 Finnish universities admitted 21,031 students, the polytechnics 25,806 students (KOTA, 2004; AMKOTA, 2004). A comparison of these numbers with the size of the relevant age cohort reveals that about 70 per cent are offered a starting place in higher education, and about that number of the young people enter higher education institutions within two or three years of their Matriculation Examination (Ahola, 2004). In 2003, there were 302,950 students in Finnish higher education institutions. Of those students 173,732 attended universities while 129,218 went to polytechnics (KOTA, 2004; AMKOTA, 2004).

The expansion of Finnish higher education can be explained by its close relationship with the goals of a welfare-state agenda supported by major political parties. The provision of equal educational opportunities became one of the most important objectives on this agenda, implemented from the 1960s to the 1990s. The expansion of higher education is consistent with egalitarian social policy found in all Nordic countries. These traditions are, in turn, historically rooted in the economic and political structures of Nordic countries with their tradition of free and independent peasants, often allied with the king against high nobility. In general, Nordic egalitarian traditions emerged in societies which have had smaller income and wealth distribution differences than other Western European countries.

In Finland the founding of a university has been seen as not only symbolically but also culturally and economically important to the development of a given region. This policy has been successful in promoting national development because it has turned out that areas with higher education institutions have been able to support knowledge-based regional economic growth, whereas other – often rural – areas are losing their labour force and economic footing. This state of affairs has also meant that those provinces which did not succeed in establishing a university have been very active in establishing polytechnics. A national policy of creating regional centres of growth has also promoted the foundation of new higher education institutions leading to the several new polytechnics in the 1990s.

Finland: Present Day

All present-day Finnish political parties share the conviction that higher education is highly important to the well-being of the country. It can also be argued that a nationalist tradition of higher education remains alive even though it has taken different forms in different periods of history. During the nineteenth and twentieth centuries, universities were an important factor in the construction of a Finnish national identity. Today, official national rhetoric sees higher education institutions as a part of the national economic innovation strategy. National policies have encouraged universities and polytechnics to cooperate with private enterprise and the world of work (Välimaa, 2001b). One indication is the fact that the proportion of funding from private sources has grown sixfold during the 1990s. However, this corresponds and is linked to steadily decreasing public funding, which began in the early 1990s when Finland was struck by a hard economic recession (see below).

In 2005, Finland had 20 universities located in all parts of the country. These include ten multi-faculty institutions, three technical universities, three schools of economics, and four art academies. All Finnish universities are state-owned public institutions in the sense of the traditional continental model. They are autonomous institutions but subordinate to the Ministry of Education. In addition to the universities Finland has 32 polytechnics distributed across the country, most of which are multi-disciplinary institutions operating under the Ministry of Education (Välimaa, 2001a). There is one higher education institution for every 100,000 inhabitants. The number of higher education institutions has been defined as one of the strengths of Finland, which aims to be the most educated nation in the world (Higher Education Policy, 1994).

Nordic Qualities and Values in Finnish Higher Education

Traditionally, Nordic higher education institutions are either state-owned universities or funded by the state, independent and autonomous institutions but regulated by national legislation and national ministries of education. Typically, Nordic higher education institutions enjoy a high degree of institutional autonomy. In Finland, this is both secured in the Finnish Constitution and guaranteed by the laws governing universities and polytechnics. It is regarded as self-evident that the institutions themselves take full responsibility for the standard and quality of the educational services they provide. The steering of the national higher education system is based on performance agreements. In this cultural context the agreements are taken seriously by the institutions and academics. In Nordic higher education, universities and polytechnics derive their formal degree-awarding capacity from the nation state. Therefore, national degrees are assumed to be comparable, equivalent and of a high standard; with the aims, scope and general structure of degrees prescribed by law (see Välimaa, 2005). There are formally no status hierarchies between higher education institutions as regards their relationship with working life. In other words, employers do not value their employees on the basis of the particular institution that granted their degree, but only on the basis of the type of education and level of degree. This is related to the fact and tradition that the state safeguards the value and quality of degrees by controlling which institutions are allowed to award them and which educational programmes qualify students for them (Välimaa, 2005; Hämäläinen et al., 2001).

It is true that among academics there may be recognised status hierarchies differentiating higher education institutions, disciplines and specialized programs of study. Students also recognise such distinctions (see Liljander, 1996). This institutional status hierarchy has not been reinforced by salary differences between individual establishments because until recently, all tenured university teachers and researchers have been civil servants with equal salaries (Välimaa 2001). This tradition is, however, challenged by an ongoing salary reform (see below).

Traditional Finnish and Nordic qualities and values are based on equality in terms of providing equal access to and gender equality in higher education, trust between higher education institutions and the state, flexibility in making changes, and the protestant work ethic with regard to the efficient use of resources (material and human) in public service. A basic question is how these traditional normative landscapes are challenged by a variety of changes taking place in Finland.

GLOBALIZATION DEBATES AND NEW CHALLENGES IN FINNISH HIGHER EDUCATION

An important aspect of present day reality in Finnish society are the current debates on globalization An interest in globalization as a social phenomenon is strong among academic researchers, and the impacts of economic globalization on Finnish society remain a sensitive issue. This sensitivity may be rooted in the economic recession of about ten years ago and in the success of Finnish IT companies like Nokia in the global marketplace after the recession. There has been also heated debates about the transfer of a number of Finnish IT production units to China in recent years and proposed large scale lay-offs in one of the nation's top international paper products company. Public interest in the topic is evidenced by the actions of Finnish political, labour and business actors relating to the local and national effects connected to the global economy.

The nature of globalization has been extensively debated in Finnish higher education in recent years. It is undeniable that higher education and research are seen as crucial for Finland's success in the global economy. This is interesting from two perspectives. First, contrary to the notion of hyper-globalization (see Held et al. 1999), the Finnish nation state continues to be an active player in the globalization game and is not willing to relinquish the authority and power associated with what many would accurately perceive as the responsibility of state government in a Nordic social democracy. Recent election results supporting champions of the Nordic welfare state bear this out quite clearly. At the same time, the elected government, and many national foundations and representatives of private enterprises, promote visions of Finland's future economic competitiveness. Secondly, higher education policy-makers proceed lock-step, stressing that education is crucial for Finland's success in the global marketplace. The high quality of universities and polytechnics is defined as critical for maintaining and improving the economic competitiveness of Finland and Finnish enterprises.

Recent policy reports advocate increasing the economic autonomy of Finnish universities to transform them into world-class universities (Brunila, 2004; Rantanen, 2004). The notion of 'world-class university' directly references to the Shanghai Jiao Tong ranking of universities, in which only the University of Helsinki was among the best 100 universities, with four others among the top 500. One specific recommendation was that only a few Finnish universities should be developed into world-class universities, while the remainder could focus on more limited national and local goals (Rantanen, 2004). These reports indicate that national resources should be channelled to research in key strategic areas (Brunila 2004). Finally, the idea has been floated that the number of both universities and polytechnics should be reduced through merger operations and increasing institutional cooperation. It has been suggested in public debates that universities should be merged with other universities and the number of polytechnics should also be reduced. Politicians advocate merger operations, speculating they seem to be a rational solution reducing higher education costs, and increasing efficiency.

These proposals are interesting in two ways. First, they indicate that global competition in higher education has been taken seriously. The politicians are, for the first time in Finnish history, not only comparing Finnish universities with universities elsewhere, but the best universities in the world. In this sense, the idea of a national higher education system is seen and defined in comparison with a highly selective group of individual universities. This is a fundamentally new basis for national higher education policy-making as it is thoroughly global in scope.

These present day debates overlook both key outcomes of present day higher education and some of the most important guiding principles of past eras. One consideration is that Finnish scholars already produce about 25 per cent more scientific articles than what would be expected given the size of the Finnish population. Another is that the doctrine of educational equality may become a casualty of 'global debate', because there is a growing consensus that international students (those coming outside of European Union) should pay tuition and fees. The rationale for this suggestion assumes that Finnish universities will improve their performance if and when they begin to compete for students in the international educational marketplace. One report even proposed that Finnish higher education could become a new 'high-tech product' in Finland (Ruokanen, 2004). What this debate studiously ignores are groups of new migrants, Finnish citizens, who may be vastly under-represented from Finnish higher education because the global nature of competition for the best students and staff consistently focuses on high profile issues. The past discussions of equity, which ironically drove policies leading to today's universal access, are fossilized in the layers of past eras. These are completely inadequate for the demographic reality of this aging and culturally homogenous population (Hoffman, 2003; Hoffman, 2006).

Public debates on the quality of academic research and the number of Finnish higher education institutions shows there is potential to reconsider the foundations of the Finnish higher education system. These suggestions in themselves indicate a challenge to the idea of Nordic higher education as rooted in the tradition of equal and free education for its citizens. These topics are not only discussed by politicians but they also have resonance inside higher education institutions, partly because they are funded by public sources, but also because of the 'global shock' experienced by Finish higher education system at the beginning of the 1990s.

"Global Shock" and Market Dynamics in Finnish Higher Education

A deep economic recession at the beginning of the 1990s was a key factor that shaped present Finnish perceptions of globalization. The economic recession meant the end of a world which had been rather predictable, given a strong commercial relationship with Soviet Russia. It also meant the end of an economic boom which had lasted for more than a decade. For Finnish higher education, economic difficulties meant a sudden end to expanding higher education budgets and deep budget cuts (Välimaa, 1994). The difficulties were so severe and indisputable that every Finnish academic saw the need to do things differently. This psychological 'global shock' prepared the ground for present day 'global dynamics' in Finnish higher education. The new feature introduced into Finnish higher education was the experience of living with increased uncertainty and under the social conditions of competition between universities and individuals. Statistical trends reveal the rapidity of the changes in Finnish higher education: the major changes in funding structures and steering mechanisms, and the effects of intensifying global dynamics in the increasing number of temporary contracts among academic personnel.

The share of public funding of higher education from the Ministry of Education decreased by 21 per cent between 1990 and 2001 (from 84 per cent to 63 per cent), while external funding from both private and public sources grew almost sixfold (KOTA 2001). Consequently, between 1990 and 1999 the number of academic teaching staff, mainly professors and lecturers, fell from 7,800 to 7,300, whereas the number of other staff on external funding almost doubled from 5,200 to 9,600 (Välimaa 2001b). According to the national KOTA database, the growth has been most rapid among staff funded by external sources (Välimaa, 2004b).

These changes reflect managerial trends associated with New Public Management (Pollit, 1995). In higher education, a strong element has been the decentralisation of management authority. Official arguments emphasise the procedural autonomy of universities in deciding how to reach the targets (the number of academic degrees) set by the Ministry of Education. According to the Ministry of Education, the steering of Finnish higher education is based on management by results. The specifics of this are that the Ministry of Education and the institutions jointly set the objectives for each higher education institution and agree on funding levels, or the number of new students in case of the polytechnics. The Ministry of Education and each university sign a performance agreement in which both parties commit themselves to production objectives measured in the numbers of degrees, development projects and level of funding. The agreement is signed for a three-year period but the financial aspects are monitored and re-negotiated every year (Higher Education Policy, 2001).

This steering instrument has introduced market or quasi-market mechanisms in Finnish higher education. The marketization of higher education has led to competition both among and within higher education institutions. The ideology underpinning the steering mechanisms in Finnish higher education is far more accurately described as academic capitalism (Slaughter & Leslie, 1997) than the ideologies or discussions associated with earlier layers of policy and practices. The shift towards academic capitalism is also changing social dynamics inside universities. A second major trend has been the requirement that staff work to performance targets and output objectives. This is closely related to a shift in the basis of public employment from permanency and standard national pay and conditions to fixed-term contracts and performance-related pay (Välimaa, 2001a).

The funding of Finnish higher education is now in crisis, compared to the situation before the 1990s. However, since the crisis is embedded in the national steering system and since new management practices are self referential, no 'crisis' exists. The fact is that only 65 per cent of the total budget of universities comes from the Ministry of Education. This funding is based on the production of MA and doctoral degrees and it barely covers the expenses of teaching staff in universities. It is now normal that additional money, to cover basic research, is only available via direct competition amongst all actors chasing those funds.

Reform of the Salary System

The reform of the civil servants' salary system implemented in the public sector for a decade as a part of the New Public Management was introduced into the universities in 2005. According to the reformers, the Finnish labour force is getting older and there is a need for performance-based salary system to make the public sector attractive to young generations. The universities are especially important here because the majority of temporary civil servant positions are in universities.

The reform is based on the idea that the individual salary of each employee is defined on the basis of two criteria; work/task requirements and personal competence and performance. The first criterion covers about two thirds of the salary, the second one third. The central problem with these aims is rooted in myths of academic productivity. During collective bargaining negotiations (see Välimaa, 2001a) representatives of the state and national labour organizations agreed on different categories for teachers, researchers and 'other staff' (meaning administrative, research support and managerial staff). At present time there are more questions than answers, including how to measure personal competence, who should do it and what criteria should be used. The main problem is how to define productivity or 'performance'. Problems are caused not only by disciplinary differences but also the difficulty of defining various categories of academic work.

The ideological starting points for this reform are based on neo-liberal thinking, specifically the rational choice theory of human behaviour, which bluntly assumes the central motive of academics centre on economic rewards. Reformers have clearly not considered the possibility that academic work is often seen as a moral profession or vocation (see Ylijoki, 2003; Välimaa, 2005b). Labour negotiators did not anticipate the spontaneous movement against the reform which emerged in the spring term 2005. About 15 per cent of the Finnish academics (more than 2500 persons) signed a web declaration insisting on an immediate stop to the process. This was an exceptional act in Finnish universities indicating an unusual degree of resistance to reform. The main arguments against the reform emphasised its tendency to create inequalities between individuals, the waste of time spent in useless discussions and the potential for intellectual control, which may be disastrous to academic freedom and independent research (Patomäki, 2005).

However, the resistance has not had much real impact on implementation of thenew system. The traditional protectors of academic staff, the labour unions, are part of the problem because traditionally, their main aim has been to improve the salaries of academic staff in universities. As such, they are full participants or partners. At the moment nobody is satisfied with the process of reform, but it will implemented in any case because the Ministry of Finances has the power to push it through. The fact that this reform can be forced in spite of concentrated opposition within academe is a new situation in Finnish higher education. The labour unions, and other parties to the process, are being resisted by a significant number of their members. How academics in Finnish higher education will deal with this blatantly global neo-liberal influence in their work is an open question.

The Bologna Process and Competitive Horizons

Another empirical insight into the globalization of higher education is provided by the Bologna Process. Its most important objectives is to enhance 'the international competitiveness of the European system of higher education institutions' (Bologna, Declaration 1999). This process initiated by European nation states aims to make European higher education one of the most lucrative higher education areas in the world through the homogenization of degree structures and European accreditation agencies (Berlin Communiqué, 2003). These tasks are quite demanding, given the variety of traditions within European higher education institutions and the national traditions of European higher education systems. For just this reason, however, it opens up an interesting perspective from which to analyse the social forces at work in European higher education and consider how globalization simultaneously influences and illuminates the functioning of universities.

The main challenge in Finland has been adaptation to changes suggested by the Bologna Process. This adaptation begun at the national level of Finnish higher education policy making in the 1990s. The challenges of adaptation are threefold: articulating and incorporating these changes in national legislation, inducing change in both the contents and structures of curricula, and creating national and institutional systems of accreditation (see Hoffman, Välimaa & Huusko, 2006).

The creation of a national system of accreditation is one of the potentially most radical changes demanded by the Bologna Process because traditionally, the Finnish system of evaluation has emphasised evaluation as a tool for development of activities. The Committee on Quality Assurance in Higher Education recommended that the Finnish higher education quality assurance system will be organized by Finnish Higher Education Evaluation Council (FINHEEC) which in turn, will organize audits in cooperation with the higher education institutions (see Välimaa, 2005). The national aim is to create a system of accreditation which is both accepted by other European nation states and which maintains the Finnish tradition of utilising evaluation as a tool for development.

Finnish higher education institutions have been rather passive actors in the Bologna Process. However, an analysis of the impact of the Bologna Process from the perspective of the basic unit reveals what is actually happening in Finnish higher education institutions, and what is not. Preparation for the implementation of reform has produced a wide variety of reactions and responses by university faculty and staff in a relatively short period of time. We will briefly summarize the central findings of studies which have analysed how personnel in basic units have defined the Bologna Process and how it has influenced unit operations.[44]

In these studies, examination of the main themes discussed in the interviews suggests three basic ways in which the Bologna Process had been translated and acted upon in basic units. The responses to Bologna can be imagined along a spectrum. On one end there are Type 1 units which have been operating

[44] The analysis is based on the following studies: Välimaa, Hoffman & Huusko (2006) and Hoffman, Huusko & Välimaa (2006). In these case studies, group interviews were done in 19 basic units in seven Finnish universities.

consistently with Bologna Process objectives for quite some time. In the middle there are Type 2 units using objectives either in instrumental or novel ways to improve their functioning. At the other end there are several Type 3 units openly resisting the very nature of the ideas and goals of the Bologna Process.

In Type 1 basic units the competitive ideology driving the Bologna Process is understood as 'the object of the game'. Personnel in these units have academic identities which are neither threatened nor challenged by this reform. This understanding explains perceptions of the reforms that range from 'business as usual' to 'a headache that was worth it' to nothing more than 'a paper exercise'. The competitive horizon of these units is global in scope. Any disturbance perceived as due to the reforms was normally translated as concrete gains for students. This understanding is rooted in the fact that the system structure the reforms are now putting in place (for example, a defined bachelors degree as a labour market qualification) corresponds more closely to the labour market their students are headed for than the pre-Bologna system had done. Some units seem very aware of this, others less so. The bottom line in Type 1 units is that the reforms - or aspects of the reforms - made a lot of sense for their students and staff.

In Type 2 basic units the Bologna Process is perceived by academics mainly instrumentally or as a novelty. In these units substantive topics or administrative reforms coming from 'outside' Finland are causing a re-think of just what constitutes 'business as usual'. The aim of increasing student mobility as suggested by the Ministry of Education, to increase new international students, programmes and subjects, provide examples of the benefits that are identified by staff in these units. In this type of unit the competitive horizon is changing, but in a ways that do not threaten or challenge the expertise, practices and identity of the academics, who seem genuinely interested in aspects of the changing global landscape.

In Type 3 basic units, the reforms introduced by the Bologna Process have no resonance with personnel because they lack relevance. The very uncomfortable and unasked question in Type 3 units was whether that lack of relevance applied to the reforms, or the participants. The most important thing about the perception and reception of the Bologna Process in these units is that there is considerable value in their criticisms. That value is not found in the (mis)perceptions of Bologna rationale or the pejorative labels participants attached to the nature of the process. The real value is attached to the idea that policy and policy measures based on international competitiveness, such as the Bologna Process or Lisbon Agenda, are only one way to estimate the relative value of work done in an academic basic unit.

We use the notion of competitive horizon to explain the variation of responses to the Bologna process in the Finnish context. To the framers of Bologna, as well as academics in Type 1 units, the competitive horizon is clearly the globe. However, the competitive horizon of academics in Type 3 units is, frankly, a hallway. The most interesting aspect, which was never articulated by the participants of these studies, is a complete reversal of the academic hierarchy. This can be seen through the historical lens we began with, in layers which are now out of living memory and completely absent from the urgent 'debates' about science policy in the age of globalization. The soft-applied academics who moulded the national debates of nineteenth and twentieth century academe into the Finnish nation state (Välimaa 2001) cannot 'compete' with the hard-applied informational economy of recent decades (see Becher & Trowler, 2001). The currency in Type 1 units is the type which can be literally deposited in a bank. However, the New Public Management has not offered the academics in the other types of units an 'exchange' rate that expresses the way in which many academics still measure themselves. (Hoffman, Välimaa & Huusko, 2006).

CURRENT PROBLEMS AND THEIR FUTURES

Funding, New Salary System and Merger Operations

As suggested above a new system-level feature of Finnish higher education reform is apparent since the 'global shock' of the 1990s. The new focus is on increasing the economic efficiency of higher education and the main instrument of the goal is competition. Neo-liberal argumentation connects the Bologna Process, the salary system reform, public debates on merger operations and discussions on the nature of higher education. The goal is not inherently in conflict with Nordic traditions, especially considering that Finland has chosen the most competitive approach to solving the problems facing all Nordic countries (see Kim, 2002). We will take a better look at the changes that have taken place before speculating on the future perspectives of Finnish higher education.

The present economic difficulties in Finnish universities, which are typical of most of the world's higher education systems, are problematic not only for higher education institutions struggling with external funding sources but also for the national system itself. The incapacity of the nation state to fund one of its most crucial factors in the global marketplace does not enhance its social credibility. The nation state's capacity to act, and the credibility connected to its actions, are likely to decrease in proportion to the amount of funding it channels to higher education.

The new salary reform, in turn, is one of the best and worst examples of the neo-liberal mechanisms introduced by the New Public Management. This reforms aims to change academic work by making it more individualistic (performance is based on personal outputs) and emphasising economic calculation as an essential motivation for academic work. It also introduces in academic basic units new social hierarchies, and reinforces existing hierarchies. It is quite possible that this reform strengthens existing competition in academia and introduces more globally accepted criteria (the most common being articles in refereed international journal) to measure one's personal performance. However, the reform is in a sharp conflict with the egalitarian traditions of Nordic higher education and the traditional civil service relationship with the state. This may be the main reason why it has been strongly resisted by Finnish academics. It is, therefore, difficult to see how much this reform really changes the nature of academic work and will this reform create new future perspectives for this traditional profession. A modest guess would be to assume that new salary system strengthens academic capitalism in Finnish universities because of its performance-based reward system.

One future possibility in Finnish higher education is the beginning of an area of merger operations in order to make existing higher education institutions more efficient. It is evident that there are too many higher education institutions, especially polytechnics, in Finland; and that some of the polytechnics are too small to be vital: 11 have less than 2900 students. The public debate about merger operations can, however, be viewed as a smoke screen which hides real funding problems in Finnish higher education. It does not seem probable that the functioning of Finnish higher institutions will be changed whether there will be more or less higher education institutions in the future. It is, however, obvious that higher education institutions feel stronger about the need to be more efficient in order to show that there is no need for merger operations (Manifest, 2006).

The Shift of Competitive Horizons in Finnish Higher Education

The most interesting aspect of the Bologna Process is, perhaps, not the fact that it is proceeding faster than anyone expected (see Lourtie, 2001; Haug & Tauch, 1999), but that the process is in fact changing the European landscape of higher education. It is changing the dynamics of national higher education systems because it has introduced alternative rationales and perspectives for the way in which academic work can be approached. In this way, the competitive horizon (of an individual, basic unit, faculty or university) empirically illuminates the quite different perspectives whereby academic personnel orientate their efforts in relation to the missions of the university. In the studies discussed above the concept of the competitive horizon is explanatory in understanding why units on the same campus perceive the purposes and potentials of the Bologna Process quite differently.

Our observations indicate emerging competitive horizons in Finnish and European higher education whether we like it or not. International indicators for comparing the capacity of a university's research programs have existed for a number of years. What the Bologna Process introduces is a structure that enables comparability with the other higher education systems, in a way that forms an instant matrix in which stratification, differentiation and ranking can occur. In addition, the new emphasis on quality assurance may provide mechanisms to simultaneously compare teaching and learning outcomes alongside research output.

Nation states also have competitive horizons. These are also changing, because of two social forces. The first competitive horizon is opened up by international comparison. Nation states need to pay attention to the comparisons like Shanghai Jiao Tong university whether they like them or not and cannot afford indifference to the potential problems they highlight (see Usher & Savino, 2005). The second important challenge is caused by quality assurance which is one of the policy objectives of the Bologna Process. Current policy says nothing about the contents of institutional quality assurance systems. It is, however, clear that a variety of institutional quality assurance systems are about to be adopted, along with audits by national and international agencies. Because the idea of quality assurance is based on standardisation, it is probable that a number of similar numerical data gathering and mining processes will need to be introduced. It is not, however, self-evident that quality assurance systems change the nature of academic activities as such, because the idea of managerially-driven quality control has always been part of academic work. However, European national accreditation systems will be ultimately judged within the global marketplace of higher education.

The Bologna Process may be leading European higher education, and Finnish universities as a part of it, towards a future where status hierarchies among universities and polytechnics are emerging. Once status hierarchies are in place, as with any form of social stratification it becomes difficult to estimate who those hierarchies benefit most, or if they in fact shift attention away from fundamentally more important debates and focal points in higher education (Bourdieu, 1988).

It is evident that international interconnectedness, globalization, has increased and will be increasing in Finnish higher education in the future. National higher education systems need to take into account global trends in the policy making processes. This does not, however, necessarily mean that nation states should absolve themselves of responsibility; either upward, in this case to the European Union, or downward to individual institutions of higher education. It does however mean that national systems of higher education can no longer be regarded as the closed social systems they have been in the past. As suggested, the competitive horizons opened by globalization have influence on the policy goals of nation states. Simultaneously, however, it is important to acknowledge that the Nordic egalitarian traditions which are part of the cultural landscape continue their influence even though economic efficiency will be emphasised more.

The elements that distinguishes Nordic higher education from everywhere else, which can be located precisely by the historical approach at the beginning of this article, are not only 'missing in action' in current debates, they may in fact be a casualty. The current debates makes very little reference to things like educational equality, the value of education as part of citizens' development (or 'bildung') and education as essential for personal growth. In other words, our vision of higher education is changing. In past eras, the people who shaped the higher education system, Finnish society, culture and nation state had more or the same horizon of the world, a national border. Today, people look at national systems using global horizons, mainly as well-off urban consumers instead of being workers or peasants.

Another change is that in past eras, the shapers of the national higher education system and nation state thought carefully about why they were doing what they were doing. Alternatives were available, discussed, even fought over. Now the perception of choices belies the fact the societies in the network in which we compete have basically all made the same choice. The currency of the present shapers of Finnish society, as much as any other OECD country, is currency. The problem is that - as Beck (1992) tries to say – we do this to ourselves because we do not think about why we are doing it. In the age of globalization it is, indeed, difficult to try to speak other possible futures for Finnish higher education.

The most important continuity with past traditions, however, is the fact that the university is understood as the national cultural institution in Finland. The fact that university academics have shaped every debate of consequence that has occurred, and is occurring in the Finnish socio-cultural context has never been disputed.

Although current problems outlined above do pose significant challenges, the venue of debate has never changed. Universities have always been important for the Finnish nation state even though the importance has been defined various ways during our history. This high appreciation can be seen in the present definitions of higher education as a part of national innovation strategy and in the confidence science and its institutions enjoy in the eyes of the general population. Universities have been crucial strongholds in the making of Finnish national identity also because they represent continuity and immaterial values, like the search for truth. The continuity of universities as cultural institutions rooted in the historical layers of Finnish society has the potential to resist and creatively channel the features of globalization that threaten the bedrock of our system. However, this can occur only as long as we are not complacent about the fragile nature of those outwardly strong layers and acknowledge that the group of people with the highest potential for destroying the most valuable features of our system are located in the same place as the only people who can ensure our system's integrity, the universities.

REFERENCES

Ahola, S. (2004). *Yhteishausta yhteisvalintaan*, 9 Helsinki: Opetusministeriön työryhmämuistioita ja selvityksiä.

AMKOTA (2004). Database maintained by the Ministry of Education, Finland. Accessed at: http://www.minedu.fi/education/polytechnics

Becher, T. & Trowler, M. (2001). *Academic tribes and territories. The intellectual enquiry and the cultures of disciplines. Second edition.* Ballmoor: Society for Research into Higher Education, UK.

Beck, U. (1992). *Risk society: Towards a new modernity.* Translated by M. Ritter. London: Sage.

Bourdieu, P. (1988). *Homo Academicus.* Translated by P. Collier. Stanford: Stanford University Press.

Brunila, A. (2004). Osaava, avautuva ja uudistuva Suomi.[Finland in the World Economy] *Valtioneuvoston kanslian julkaisuja*, 19/2004. Helsinki: Edita.

Finnish Science Barometer (2004). Finnish Society for Scientific Information. Helsinki: Yliopistopaino.

Hämäläinen, K., Haakstad, J., Kangasniemi, J., Lindeberg, T., Sjölund, M. (2001). *Quality assurance in Nordic Higher Education.* ENQA Occasional Papers, 2. Helsinki.

Held, D., McGrew, A., Goldblatt, D. & Perraton, J. (1999). *Global transformations: Politics, economics and culture.* Stanford: Stanford University Press.

Higher Education Policy (1994). Ministry of Education. Helsinki.

Higher Education Policy (2001). Ministry of Education. Helsinki.

Haug, G. & Tauch, C. (1999). Toward *the European higher education area: Survey of main reforms from Bologna to Prague.* Accessed at: http://www.bologna-berlin2003.de/pdf/trend_II.pdf

Hoffman, D. (2004). 300 Meters. In R. Koikkalainen (Ed.) *Koettu Kampus*, 53-80. Saarijärvi, Gummerus.

Hoffman, D. (2006). *Social responsibility, educational equality and Finnish higher education.* In J.Ursin & J. Välimaa (eds.) Kritiikkiä, teoriaa ja keskustelua korkeakoulutuksesta [Theories and critical debates on Higher Education].

Hoffman, D., Huusko, M. & Välimaa, J. (2006). The competitive horizons of academic basic units in Finnish universities in the context of the Bologna Process. Consortium of Higher Education Researchers 18th annual conference, Jyväskylä, September 1-3, 2005 [revised].

Kim, L. (2002). Lika olika. En jämforelse av högre utbildning och forskning i de nordiska länderna. *Högskoleverkets rapportserie*, 40 R. Stockholm: Högskoleverket.

KOTA (2004). Database maintened by the Ministry of Education, Finland. Accessed at: http://www.csc.fi/kota

Liljander, J.-P. (1996). *Statusvoittojen ja –tappioiden tiet korkeakoulutuksessa.Travelling the roads of status gains and losses in higher education*. Jyväskylä: Institute for Educational Research.

Lourtie, P. (2001). *Furthering the Bologna process. Report to the Ministers of Education of the signatory countries.*

Manifest published by Finnish Council of University rectors (2006). Accessed at: http://www.rectors-council.helsinki.fi/

Nevala, A. (1999). Korkeakoulutuksen kasvu, lohkoutuminen ja eriarvoisuus Suomessa [Growth, Fragmentation and Inequality in Higher Education in Finland]. *Bibliotheca Historica*, 43. Helsinki.

Rantanen, J. (2004). Yliopistojen ja Ammattikorkeakoulujen tutkimuksen rakenneselvitys. *Opetusministeriön työryhmämuistioita ja selvityksiä*, 36.

Ruokanen, T. (2004). *Suomen Menestyksen eväät. Tiekartta tulevaisuuteen.* [Finland's Roadmap to Future] Helsinki: Taloustieto.

Slaughter, S. & Leslie, L.L. (1997). *Academic capitalism: Politics, policies and the entrepreneurial university*. Baltimore: Johns Hopkins University Press.

Usher, A. & Savino, M. (2006). A world of difference: A global survey of university league tables. Accessed at: http://ww.educationalpolicy.org

Välimaa, J. (1994). A trying game: Experiments and reforms in Finnish higher education, *European Journal of Education*, 29(2), 149-163.

Välimaa, J. (2001a). A historical introduction to Finnish higher education. In J. Välimaa (Ed.) *Finnish higher education in Transition – Perspectives on massification and globalization*. Jyväskylä: Institute for Educational Research, 13-54.

Välimaa, J. (2001b). The changing nature of academic employment in Finnish higher education. In J. Enders (Ed.) *Academic staff in Europe: Changing contexts and conditions*, 67-90. Westport, Conn.: Greenwood.

Välimaa, J. (2004b). Nationalisation, localisation and globalization in Finnish higher education. *Higher Education*, 48 (1), 27-54.

Välimaa, J. (2005). Globalization in the concept of Nordic higher education. In A. Arimoto, F. Huang, & K. Yokoyama (Eds.) *Globalization and higher education*. RIHE International Publication Series 9, 93-114. Hiroshima University: Hiroshima.

Välimaa, J. (2005b). Akateeminen työ –palkkatyötä vai säätyläisyytä? [Two Dimensions of Academic Work]. In O.-H. Ylijoki & H. Aittola, H. (Eds.) *Tulosohjattua autonomiaa –Akateemisen työn muuttuvat käytännöt*, 146-168. Helsinki: Gaudeamus.

Välimaa, J, Hoffman, D., & Huusko, M. (2006). The Bologna process: Perspectives from basic units. In V. Tomusk (Ed.) Creating The European Area of Higher Education. Voices from the Periphery. Dordrecht: Springer, 43-68.

Ylijoki, O.-Y.H. (2003). Entangled in academic capitalism? A case-study on changing ideals and practices of university research. *Higher Education*, 45, 307-335.

Jussi Välimaa,
Institute for Educational Research
University of Jyväskylä

David M. Hoffman
Institute for Educational Research
University of Jyväskylä

MICHAEL PETERS & TIM MAY

10. HIGHER EDUCATION, REGIONAL POLICY AND THE KNOWLEDGE ECONOMY

The North-West of the UK

INTRODUCTION: THE NEW ECONOMY?

In the West educational policy has given way to talk of the knowledge economy under the sway of world policy organizations like the Organization for Economic Co-operation and Development (OECD) and the World Bank. The terms 'knowledge economy' and 'knowledge capitalism' emerged in the mid-1990s to become national policy templates for many Western governments and developing economies. The United Kingdom (UK) Government, for example, has pronounced the end of the comprehensive school, based on a 'one-size-fits-all' welfare-state ideology, and signalled a shift to a fully consumer-driven system of public services in health and education, based on the market ideology of choice and diversity. This shift to the 'social market' is underwritten and accompanied by an emphasis on national competitiveness within the global economy and the way in which the 'new economy' demands new levels of flexible skilled knowledge workers. This chapter examines the discourse of the knowledge economy and education as an 'industry of the future' that can promote regional development, with the attendant emphasis on public–private partnerships and the cultural reconstruction of city entrepreneurial cultures and clustering of knowledge capital activities.

Digitalisation, speed and compression are the forces at work that have transformed the global economy and now have begun to affect every aspect of knowledge production – its organization, storage, retrieval, and transmission. The knowledge economy has certainly arrived, although this does not mean the end of the business cycle, as many early advocates of the new economy maintained. But it does signal structural economic shifts and new sources of growth in some Western economies (for example USA, Finland) that delivered both low unemployment and low inflation due to increased productivity. While it is clear that investment in information and communications technology (ICT) and ICT-driven productivity growth has led to a higher growth path, there is a risk of exaggerating the growth potential due to ICT investment alone. Yet as a recent OECD report put it:

> It would be wrong to conclude that there was nothing exceptional about the recent US experience, that the new economy was in fact a myth. Some of the arguments posited by new economy sceptics are of course true: the effect of ICT may be no greater than other important inventions of the past, like electricity generation and the internal combustion engine. Moreover, far

Simon Marginson (ed.), Prospects of Higher Education, 201–213.

> greater productivity surges were recorded in previous decades, not least in the period before the 1970s (OECD, 2001, p. 10).

Nevertheless, the evidence suggests that something new is taking place in the structure of OECD economies (OECD, 2001, p. 10). The report continues by maintaining that ICT has facilitated

> productivity enhancing changes in the firm, in both new and traditional industries, but only when accompanied with greater skills and changes in the organization of work. Consequently, policies that engage ICT, human capital, innovation and entrepreneurship in the growth process, alongside fundamental policies to control inflation and instil competition, while controlling public finances are likely to bear the most fruit over the longer term. (p. 10)

Crucially, the report investigates and recommends a set of relationships and policies that harness ICT, human capital, innovation and business creation, focusing on the wider diffusion of ICT and the role of education and training policies in meeting today's skill requirements.

The Internet economy is becoming an integral part of the global economy, creating jobs, increasing productivity and transforming companies and institutions. Employment in the Internet economy is growing faster than in the traditional economy. In the US economy alone the Internet generated an estimated $830 billion in revenues in 2000, which represented a 58 per cent increase over 1989.[45] J. Bradford De Long, former Deputy Assistant Secretary for Economic Policy in the US Department of the Treasury, depicts the 'new economy', asserting it is both a knowledge and an innovation knowledge where clusters of innovation, based on new technologies and new business models, succeed each other. He maintains it is likely to continue for an extended time with pervasive consequences. He provides an analytical overview of the digital economy which conveys how different it is from the market economy of orthodox economics. He likens the digital economy to the enclosure of the common lands in early modern Britain, which paved the way for the agricultural and industrial revolutions. Digital commodities, he maintains, do not behave like standard goods and services of economic theory: they are non-rivalrous, barely excludable and not transparent. The store of music tracks is not diminished when one downloads a track from the Internet; it is difficult, if not impossible, to restrict distribution of goods that can be reproduced with no or little cost, and; a consumer does not know how good software is before purchase or indeed how its successor versions will perform in the future.

The knowledge economy is both classical and new. Danny Quah of the London School of Economics indicates that the economic importance of knowledge can be found in examples where deployment of machines boosted economic performance such as in the Industrial Revolution. By contrast, he talks of the 'weightless economy', 'where the economic significance of knowledge achieves its greatest contemporary resonance' and suggests it comprises four main elements:

---

[45] See The Internet Economy Indicators at http://www.internetindicators.com/execsummry.html

1. Information and communications technology (ICT), the Internet.
2. Intellectual assets: Not only patents and copyrights but also, more broadly, name brands, trademarks, advertising, financial and consulting services, and education.
3. Electronic libraries and databases: including new media, video entertainment, and broadcasting.
4. Biotechnology: carbon-based libraries and databases, pharmaceuticals. (http://econ.lse.ac.uk/staff/dquah/tweirl0.html)

Elsewhere he argues:

> Digital goods are bitstrings, sequences of 0s and 1s, that have economic value. They are distinguished from other goods by five characteristics: digital goods are nonrival, infinitely expansible, discrete, aspatial, and recombinant. (Quah, 2003, http://econ.lse.ac.uk/staff/dquah/dp-0212hbne.html)

Quah (2001) also has been influential in suggesting that knowledge concentrations spontaneously emerge in space, even when physical distance and transportation costs are irrelevant. The dynamics of spatial distributions manifest themselves in convergent clusters. This is an important feature, especially given the development of the e-conomy first in Silicon Valley.

Coyle and Quah (2002), while noting the 11 per cent productivity gap between the UK and Germany, and the 45 per cent productivity gap between the UK and the USA (1999 base year), at the same time acknowledge the considerable impact of ICT on the economy. They demonstrate on the basis of evidence from the USA that those businesses responding the quickest to ICT developments are driving forward the rest of the economy. At the same time they maintain that technology takes time to filter through and set up the cascade effects that are evident in change organizational and business practices. They acknowledge that after the dot-com bubble burst, confidence in the new economy has waned yet they remain optimistic about long-term, technology-led economic growth based on the processing power of the microchip, which encompasses well-known developments like the Internet and developments in information and communications technologies with gene technology, nanotechnology, robotics and advanced materials. They write:

> Advocates of the notion of the new economy cited supporting factors uniquely combined in the US economy, particularly minimal government, high levels of competition, encouragement of entrepreneurship, and access to venture capital. What gave the theory bite, though, was the vision that the widespread diffusion of technology had permanently changed the way economies had worked for the better (Coyle & Quah, 2002)

They also develop a set of indicators to measure the changes to the deep structure of the economy brought about by technology.

CLUSTERING EFFECTS OF NEW ICT-DRIVEN ECONOMIC ACTIVITY

One of the most significant features of these economic changes is the way in which against expectations a sort of geographical clustering of new economic activity appears to have taken place. This clustering even applies to the broader accounts of its development and spread. The Silicon Valley view of the economy maintains that the centre–periphery model, with development spreading from the West Coast, south of Palo Alto, forced the development of a single international market that is dominated by American multinationals. But the pluri-view suggests that Internet leading-edge use and ecommerce centres are beginning to emerge in Europe and Asia, and as new technologies emerge, along with different uses, business models and legal frameworks, they will challenge the early dominance of American policy and the international market. If there are different local configurations of market demand and distinct trajectories of development and separate national economies, they are most likely to emerge around fundamental empirical criteria such as distinct technologies and applications; locally differentiated market structures; different business models; perhaps, distinct structures of comparative advantage; and culturally different legal and policy frameworks.

The alleged aspatial character of digital goods seems to fly in the face of empirical evidence. Given their 'weightlessness', digital goods are theorised to spread easily across the globe without favouring particular locations, yet empirical analysis reveals exactly the opposite – that the production of digital goods like other goods becomes geographically concentrated. As Quah (2002, p. 38) argues: 'the geographical clustering of computer software and digital media production, academic and commercial R&D, and financial services, among other digital goods, is likely tighter than for ordinary goods and services'. Further, he states, the aspatiality of digital goods does not imply that space no longer matters. Only transportation costs no longer matter, others factors associated with geography take on a heightened importance. He indicates that embodied human capital clusters geographically because communication of tacit knowledge is most efficient in close proximity. In other words, 'synchronous face-to-face interactions matter for transmitting (nonbitstring) knowledge' (Quah, 2002, p. 39). On this understanding sometimes learning is considered the core of a theory of clustering, focusing on the existence, the internal organization and boundaries of the cluster (Maskell, 2001).

The major theoretical question is why has industry clustering reappeared in advanced economies when it had all but disappeared in the mid-twentieth century. Cooke (2002) suggests that the knowledge economy consists in fragments and runs against conventional economic analysis that assumes individualistic competition within an ordered economic equilibrium. By contrast, with the knowledge economy he points to disequilibrium or economic and social imbalance, collaborative economic action as the basis of modern capitalism, and the systemic nature of strategic competitiveness of groups' actions based on consensus rather than individual opportunism (Cooke, 2002, p. 1). Against the status of the lonely Nietzschean heroic individual and entrepreneur characterised by Schumpeter, Cook suggests the sources of innovation lie in networks of social relationships, that is, knowledge networks. He argues:

> clusters are crucial to economic imbalance, ... they rest upon collaboration of a generally non-market-destroying type that is simply essential for modern economic organization, and ... clusters have systemic organizational characteristics that go against much economic orthodoxy (Cook, 2002, p. 2).

Such an analysis then directly leads to an emphasis on 'the geographical dimension of learning, knowledge transfer, collaboration and the exploitation of the spillovers' (p. 2). Cook concurs with Quah that 'proximity in a cluster offers the opportunity for tacit knowledge exchange or "treacherous" learning that may be hindered in large firms by "group think" and corporate culture' (p. 3).

Government policy has changed to accent and foster the geographical cluster of new economic activity. It is clear that that Labour Government sees the 'knowledge-driven economy' as key to the UK's competitive success. The strategy behind the adoption of this concept is to promote high-tech venture capital 'clusters' in the regions. In part the UK strategy is strongly influenced by Michael Porter's (1998) work on the role of clusters in forging US innovativeness and competitiveness. The regional development agencies (RDAs) in the UK are encouraged to develop 'cluster' policies based on knowledge industries. For instance, the White paper on *Enterprise, skills and innovation* (Department of Trade and Industry, 2001) emphasises the role of universities in the regional development of clusters in the knowledge economy:

> The role of our universities in the economy is crucial. They are powerful drivers of innovation and change in science and technology, the arts, humanities, design and other creative disciplines. They produce people with knowledge and skills; they generate new knowledge and import it from diverse sources; and they apply knowledge in a range of environments. They are also the seedbed for new industries, products and services and are at the hub of business networks and industrial clusters of the knowledge economy.

The examples can be developed further. The UK Office of Science and Technology (OST) emphasises the importance of 'knowledge transfer':

> Within a modern, knowledge driven economy, knowledge transfer is about transferring good ideas, research results and skills between universities, other research organizations, business and the wider community to enable innovative new products and services to be developed (OST, 2004).

This emphasis is clearly seen in OST's Higher Education Innovation Fund (HEIF):

> We have now established the Higher Education Innovation Fund (HEIF), which together with the existing Higher Education Reach Out to Business and the Community fund (HEROBC), is worth £140 million over three years, to further build on universities' potential to act as drivers of growth in the knowledge economy. HEIF provides special funding to enable HEIs to respond to the needs of business, through both the continuing development of capacity in universities to interact with business and the community (building

on HEROBC), and large, strategic, collaborative projects to strengthen university–business partnerships. (OST, 2004)

More broadly, the UK Government's economic strategy can be seen as being based on investment in science, engineering and technology. In the recent White paper *Investing in innovation: A strategy for science, engineering and technology* (OST, 2002) the focus is on the twin strategies of renewing the physical and human capital which underpins research, and investing in capacity to exploit the pportunities of new science. In this strategy universities are encouraged to link with business to create value for the regional and national economy where RDAs play an enhanced role in developing knowledge transfer programmes.

The regional development role of knowledge industries has received increasing attention in the USA and by the European Union (EU). CEOs for Cities, for example, an organization that explores best practices in urban economic development and proposes new strategic frameworks on emerging trends, recently partnered with the Initiative for a Competitive Inner City to develop a national study of the impact of higher education on urban economies. The study introduces a strategic impact framework and features case studies and best practices in university partnerships, including studies of both Virginia Commonwealth University in Richmond and Columbia University in New York City. The report begins:

> Unleashing the local economic development capacity of these institutions (colleges and universities) should be a national priority. While ambitious, it is an agenda that does not require massive new funding or heroic changes in day-to-day operations of colleges and universities, city government or community groups. (Joint Study by Initiative for a Competitive Inner City and CEOs for Cities, 2002, p. 2)

The Lisbon European Council set the 'bold and ambitious' 10-year goal of making the EU the most dynamic, competitive, sustainable, knowledge-based economy in the world. Yet the transition to the knowledge-based economy has been slow and there is now recognition that spending on higher education needs to be strengthened. The argument is made that the European ability to produce, diffuse and use knowledge effectively relies heavily on its capacity to produce highly educated people for its firms to be engaged in a continuing process of innovation.

Yet lifelong learning is still not a reality for most European citizens. It is now recognised that the European innovation systems have not been successful in exploiting the new techno-economic paradigm rooted in information technology that is reflected in the European paradox of a strong science base but weak innovation performance. As Lundvall & Borras (1998) argue: 'A major policy objective of innovation policy must be to contribute to the learning capability of firms, knowledge institutions and people and to promote innovation and adaptation'. They point to 'human resource development, new forms of firm organization, network formation, new roles for knowledge intensive business services and for universities as the key elements in speeding up the catching-up within this paradigm'. A new regional policy focusing increasingly on the role of universities and of higher education more generally, first recognised in the Bologna agreement, has begun to theorise these economic imperatives in new spatialised knowledge networks. The role of universities in regional development has gone beyond the study of technology transfer and direct employment effects of spin-off companies and the establishment of science parks, to embrace the wider ethos of the enhancement of human and social capital within a region, including emphasis on student recruitment and regional placement policies, university professional development programmes for local managers based on local research, the development of research networks that serve as a basis for embedding the local businesses in the global economy, and a more research and information-sensitive negotiation of the local/global interface.

UNIVERCITIES: THE MANCHESTER KNOWLEDGE CAPITAL INITIATIVE

In the Northwest of the UK, this policy context is reflected in several recent developments. The North West Universities Association was formed in 1999 and works increasingly closely with the Northwest Development Agency (NWDA) and other regional partners. In 2001, the North-west established England's first science council, bringing together representatives from industry, regional agencies and the universities to lobby on behalf of the region and advise and launch the Regional Science Strategy (NWDA, 2002). This was subsequently published in 2002 and sets forward cluster-based actions in five priority areas (biotechnology, environmental technologies, chemicals, aerospace and nuclear energy) to link universities better with industry and regional partners. The model is one of 'excellence with relevance'.

Many regional development agencies are now devolving the implementation of aspects of regional strategies to sub-regional bodies, recognising that some issues are better tackled at a lower spatial scale. At the same time, local authorities are collaborating with neighbours and partners to 'upscale' their cities and tackle joint issues through a cross-boundary approach. The movement also now includes health authorities, universities, local education authorities, skills agencies, charities and the police who are active in their own city-regional partnerships. In a number of cases the private sector is giving the lead to city-regional thinking, particularly in the arena of economic and planning policy (Centre for Sustainable Urban and Regional Futures, 2003). In the Greater Manchester context, we see the development of the Greater Manchester sub-regional strategy, led by the Association of Greater Manchester Authorities (AGMA), launched on 20 June 2003. An essential element of this strategy focuses on enhancing the sub-regional core and building on the university assets concentrated there (AGMA, 2002, p. 16).

With this in mind the vision of 'Manchester: Knowledge Capital' (KC) is:

> to create an internationally acclaimed 'Knowledge Capital' within the Greater Manchester conurbation, which will position Manchester, branded as the Knowledge Capital, at the heart of the Knowledge Economy, significantly contributing to the economic growth of the nation and the North West region leading to a healthier city/region with a vibrant, safe and attractive

> environment in which to live, work and play, for people of all ages, social and cultural backgrounds. (http://www.manchesterknowledge.com/knowledge. html).

This initiative can be seen in the context of two major changes fundamentally reshaping our society, economy and the institutions and practices of modern life: the development of the knowledge economy and changing notions of scale. These two factors have led to increasing attention being given to the role of universities in driving local and regional economies, reflected in a wealth of new policies being developed at international, national, regional and sub-regional scales. These policies raise many challenges for universities and for their localities and a number of different responses, as international and national comparisons indicate. In this context, KC represents a much needed and valuable opportunity for the universities to position themselves at the heart of the city-region's response to the global knowledge economy.

From the 40 interviews and documentary work conducted in the four universities (University of Manchester, Manchester Metropolitan University [MMU], University of Salford and UMIST) with senior managers and those individuals who were considered key to successful implementation of KC, we can see the extent to which understandings remain aspirational at a variety of levels across the universities. Relatively speaking, the most well-developed attempts at clarification of the meaning of KC are provided, not surprisingly, by those who are championing and driving the process for the universities. Overall, these are a small number of individuals in senior management positions. Here we see an explicit identification of a role; that is, creating the conditions that will enable its development. The role of senior managers is to create not only a vision, but also an infrastructure that will function without too much top-down planning to provide support and incentives.

Yet a clear aim of KC and one given by central government is that the universities should act in collaboration with each other and with other agencies for the benefit of the city-region. This was both implicit and explicit in many of the interviews. In particular, one senior manager spoke of a great strength of the city of Manchester being a 'mutual trust between the key players'. This is matched with the view of one senior manager in Manchester City Council who said that KC provided a context in which: 'we construct the new HEI sector in such a way that it interfaces very effectively and coherently with a significantly wider world [than has been historically the case]'. Many of these interviewees whilst unpacking, or alternatively trying hard to populate, the notion of KC also sought clarification as to its geographical focus. Some took it to concentrate on the North-west of England, others Greater Manchester, some Manchester, but all with a degree of uncertainty. The idea of 'capital' itself was also a source of ambiguity, with ideas of human capital and innovation mixing with those of cultural and physical capital, and Manchester seeking to place itself more securely as the 'capital' of the North-west. Importantly, what we see here are differences in aspiration and interpretation according to the remit of the individual interviewed. Those further up the hierarchy tended to be supportive of the idea in terms of its potential, whilst those with a concern for implementation and its implications for practice tended to be more uncertain. This belief came from an experience of so many 'next big thing' initiatives that produced more work without tangible benefit. It also came from the institutional incentives that were available for engagement at different levels. As one interviewee put it in relation to practices surrounding international research excellence (of which they were a part), KC should be about:

> achieving some sort of culture shift within the institution ... the notion that you can only be taken seriously if you go to conferences in North America or Munich or whatever and that anything where you dirty your hands on things local, I mean, by definition is seen as trivial ... the international and the local can readily co-exist together ... [the University needs] levers to pull locally and to ensure that this might happen and all of the incentives and reward structures still tend to reinforce the argument.

Incentivisation and reward within the different institutions and how they relate to practice is clearly a major issue and one that has not been adequately considered in relation to the developmental potential of KC.

Two issues then become of importance. First, to develop the conditions that facilitate activities and second, for those activities to feed into KC in order that it moves from concept to action. What is required for this process is a set of practices that would populate the concept. As one person put it, it cannot just be about:

> museums and office blocks and such like in the centre, but actually focus on regeneration through knowledge applied to entrepreneurial activities ... that seems to me to be an excellent extension and the next stage after we've physically transformed the city into something worth living in ... then we've got to jump start some entrepreneurial activity or it will be another Sheffield Centre for Popular Culture!

There were also differences between the institutions in terms of how they approach the potential of knowledge capital. Those associated with UMIST/Victoria tended to see the development of KC as one of 'added value'. It was concerned with repackaging and exploiting existing opportunities for institutional benefit and that of the city as a whole, in terms of the location of the University and its relations with key stakeholders. This concerned the development of relations of mutuality with the City Council and other organizations in providing a context and environment for attracting staff, finance, investment and facilities to Manchester. It would also provide further evidence of an 'innovative milieu' through the development of incubation facilities for the exploitation of knowledge, infrastructure (physical and human) and other visible signs of activity: for example, cultural in relation to art galleries, theatres and museums. To this extent the development of KC is part of the overall strategy of the 'Project Unity' merger in seeking to counterbalance the 'Golden Triangle' Oxford, Cambridge and London, becoming a 'Harvard of the North' and part of an extended 'Golden Diamond'.

Such international aspirations should be seen against a change in recent years in which both institutions have moved a long way from being seen as 'in' but not 'of' the region. It was emphasised that the shift towards an internationally focused and renowned institution was not incompatible with a local and engaged focus as it was a contributor to the 'well-being' of an area. The development of Victoria as a world-class research university was illustrated by developments in bio-medical research (a £40 million complex for bio-medical research relates to what was described as a 'bio-medical corridor' in Oxford Road) and growing relationships between medical facilities, the National Health Service, pharmaceutical companies and spin-out enterprises. Victoria's focus was in positioning itself as an international, first-class, science-based, research university able to attract the brightest students and leading edge academics and develop the facilities that match these aspirations to particular developments. The associations that exist between academics within Greater Manchester and the North-west region temper the international dimension of activities in some departments: for example, bio-medical science and collaborative links with Liverpool. The level of activity may not directly relate to the locality. That is not to suggest it lacks local implications. For instance, in terms of the opportunities presented by the merger for new centres of research excellence, this person spoke about a centre for climate change:

> the atmospheric physicists know all about modelling of the climate and atmosphere etc, making connections with the earth sciences, setting up a more environmentally orientated physics and earth sciences school ... the merger is an opportunity to reconfigure into these new knowledge lumps ... so in that sense there's tremendous opportunity for rearranging some of the intellectual furniture, and I would have thought that presents opportunities for new research directions focused on spin off and entrepreneurial activities.

Allowing such possibilities to flourish over time, as opposed to imposing a vision upon sets of practices, is precisely how some of the most advantageous developments may take place. In addition, whilst there are international aspirations, there are also different levels of engagement within Victoria and UMIST. The departments of Planning and Landscape, and Architecture, for example, have worked on KC design projects and possess a long tradition of working with local communities in terms of outreach, widening participation and issues associated with multiculturalism.

KC at MMU tended not to be seen as a repackaging of existing processes, but as a means of continually cultivating relationships between the University and the region around a number of strategic themes in which they were investing their energies and resources. These areas of activity were: Network for Enterprise; Innovation in Art and Design; Regeneration; Sport and Physical Activity; Clothing, Design and Technology and Aviation, Transport and Environment.

KC was regarded as enhancing those areas of activity, as well as being transformative. In the process it brought greater recognition to the University in terms of its overall identity. This has been prevalent in terms of creativity and culture, for example, art, design, fashion and sport. MMU was regarded as having a strong vocational base with a regional focus. The University possessed clear areas of research excellence (seven four-star departments and one five-star department in the latest Research Assessment Exercise), but was seen as having a focus on widening participation through outreach activities, through enterprise in local schools, through the large number of teachers trained in the institution, continuing professional development and via initiatives like the Community Entrepreneurship Scheme. Problem solving and innovation with local businesses led in interviews to an emphasis upon practical engagement. For example, in terms of fashion, MMU deals with developments and ideas for clothes which retail rather than high fashion. Thinking about this emphasis in terms of seeking to welcome people into universities who would not otherwise come in, this person noted in respect to the Manchester Fashion Network that there was an opportunity to enhance recognition through KC. It was viewed as an important means of providing coordination and coherence to a wide variety of activities that saw the University seeking to reach out to people who would not normally engage with higher education.

This latter aspiration is shared with Salford, in which one senior manager viewed KC as an 'infrastructure of possibilities'. Here academics and the institutions could form relationships of knowledge creation, production and sharing for multiple beneficiaries. KC was an aspiration to create a structure within which people can move and be creative. There was also the sense generally that KC related to aspirations in terms of the mixed aims at Salford of teaching, research and enterprise. The flow of knowledge and the movement of people within a defined area was key to this notion.

Salford was seen to have notable pockets of research excellence: for example, in Information Systems, the Built Environment, European Studies and Urban Regeneration, Media, Art and Design, Public Health, as well training in relation to Professions Allied to Medicine. Its reputation in these areas was seen in terms of its distinctive strengths in addressing business, industrial and commercial interests in, for example, the design and deployment of 'enabling technologies' through working in partnership. This ethos was linked to the potential seen in KC. An emphasis was also placed on widening access and participation via a number of initiatives, with the aim of raising young people's aspirations. Mentoring was, for example, one programme of work mentioned, whilst the relationship between skills and student was seen to be accommodated within Salford. Overall, therefore, KC was seen as enhancing existing agendas in teaching, enterprise and research, all of which were driven by the idea of being an 'engaged' university working in partnership with a variety of stakeholders.

All this takes place against a background of change in which understanding differences between institutions, as well as similarities, is required. At the launch of the Higher Education Funding Council for England's draft strategic plan (2003-08), Sir Howard Newby was quoted as saying that individual universities: 'must build upon their own chosen areas of strength, and work in collaboration with other providers, so that the sector as a whole continues to deliver all that is required of it in the increasingly competitive global marketplace'.

Complementarity, on the basis of an understanding of distinctiveness, was seen as crucial to the development of KC by all those interviewed. This is not to suggest that there is not competition between the universities as they seek to differentiate their niche markets and convey a particular identity to an outside audience. That means recognising the distinctive strengths of each institution.

In terms of further engagement and making the universities meaningful to citizens, a large number of deprived wards surround Manchester. As one person put it: 'we have a huge mountain to climb in terms of raising the aspirations of local people ... young people'. Spatially speaking, this also relates to a tension between what is seen as the 'urban core'. KC might be very successful, for example, in attracting a larger pool of knowledge workers. Some of these may stay in the 'centre', but there are no schools and so where will they go if they have, or want, children? In this sense, how will this community benefit those who already exist in areas such as East Manchester, Hulme and Moss Side? How are these latter groups to benefit from KC?

New urban and regional education futures have recently become the basis for both the knowledge economy and also dreams of the renewal of the post-industrial city. This movement that depends upon a reconceptualisation of regional development has the prospect of emphasising the university as a local source of research, expertise and innovation and providing a knowledge base for local industries to thrive in the global economy. Perhaps, more importantly, it holds the promise of rearticulating the links between the university and the communities it serves to enhance civic traditions and thickening democracy.

REFERENCES

Association of Greater Manchester Authorities, AGMA (2002). Draft Greater Manchester strategy. Wigan: AGMA.

Centre for Sustainable Urban and Regional Futures, SURF (2003). Evaluating urban futures: Enhancing quality and improving effectiveness. SURF, University of Salford. Accessed at: http://www.surf.salford.ac.uk

Cooke, P. (2002). Knowledge economies: Clusters, learning and cooperative advantage. London: Routledge.

Coyle, D. & Quah, D. (2002). *Getting the measure of the new economy*. Accessed at: http://www.theworkfoundation.com/pdf/1843730014.pdf

Department of Trade and Industry (2001). Opportunity for all in a world of change: A white paper on enterprise, skills and innovation. Accessed at: http://www.dti.gov.uk/opportunityforall/

Joint study by Initiative for a Competitive Inner City and CEOs for Cities (2002).

Leveraging colleges and universities for urban economic revitalization: An action agenda. Accessed at: http://www.ceosforcities.org/research/2002/leveraging_colleges/

Lundvall, B.-A. & Borras, S. (1998). Innovation policy in the globalizing learning economy. Accessed at: http://www.cordis.lu/tser/src/sumfinal.htm

Maskell, P. (2001). Growth and territorial configuration of economic activity. Accessed at: www.druid.dk/conferences/nw/paper1/maskell.pdf

Northwest Development Agency, NWDA (2002). Science strategy England' s Northwest. Warrington: NWDA.

Office of Science and Technology, OST (2004). Knowledge transfer', 'Exploitation funding' webpage. Accessed at: http://www.ost.gov.uk/enterprise/knowledge/index.htm

Office of Science and Technology, OST (2002). Investing in innovation: A strategy for science, engineering and technology. Accessed at: http://www.ost.gov.uk/policy/science_strategy.pdf

Organization for Economic Cooperation and Development, OECD (2002). The new economy: Beyond the hype. Accessed at: (http://www.oecd.org/dataoecd/2/262380634.pdf)

Porter, M. (2000). Location, competition and economic development: Local clusters in a global economy. Economic Development Quarterly, 14(1), 15-34.

Quah, D. (2001). Demand-driven knowledge cluster in a weightless economy. Accessed at: http://econ.lse.ac.uk/dquah/p/0104dkc.pdf

Quah, D. (2002). Digital goods and the new economy. Centre for Economic Performance, London School of Economics. Accessed at: http://cep.lse.ac.uk/pubs/download/dp0563.pdf

Quah, D. (2003). Digital goods and the new economy. In D.C. Jones (Ed.) New economy handbook. New York: Academic Press.

Michael Peters
Department of Educational Policy Studies
University Of Illinois (Urbana-Champaign), United States of America

Tim May
Centre for Sustainable Urban and Regional Futures
University of Salford, Manchester, United Kingdom

RAJANI NAIDOO

11. FIELDS AND INSTITUTIONAL STRATEGY:

Bourdieu and the Relationship between Higher Education, Inequality and Society in South Africa

INTRODUCTION

At the heart of Bourdieu's work on higher education has been his desire to expose higher education as a powerful contributor to the maintenance and reproduction of social inequality. This overarching focus has led to a significant theoretical and empirical contribution to sociological understandings of the relationship between universities and society. Bourdieu has attempted to achieve an 'unveiling' of the functioning of higher education through his development of the core concepts of 'field', 'capital' and 'habitus'.

While these constructs function fully only in relation to one another, this chapter[46] has, as its focus, a critical exploration of the concept of 'field'. It begins by presenting Bourdieu's theoretical framework to examine how the concept captures the sense of higher education as a force that mediates, and at the same time reproduces, fundamental principles of social classification. As an illustration of its widespread application, the framework is applied to develop an analytical understanding of divergent admission policies developed by South African universities during a period of political instability. Drawing insights from this study and Bourdieu's empirical research, the strengths of the field concept are outlined. In the final section of the chapter, limitations as well as directions for further development are presented.

THE FIELD OF HIGHER EDUCATION

According to Bourdieu, social formations are structured around a complex ensemble of social fields[47] in which various forms of power circulate. The relative

[46] The author thanks Madeleine Arnot and Joe Muller for their support and guidance in the development of the empirical work on which this chapter is based; and Hugh Lauder, Ian Jamieson, Graham Thom and two anonymous referees for their helpful comments on earlier versions of the work.

[47] The differentiation of the social world from a state of non-division, regarded as characterising primitive society, to the separation of diverse social functions in more complex society, for example, scientific thought from religious thought, and art from worship, has previously been analysed by Durkheim. For an illustration of how Bourdieu draws upon and extends the Durkheim's thesis, see Bourdieu & Wacquant, 1992, pp.12-14).

Simon Marginson (ed.), Prospects of Higher Education, 215–228.

autonomy of fields varies from one period to another, from one field to another and from one national tradition to another (Bourdieu, 1993). In much of Bourdieu's research and the work of others drawing on his framework (see, for example, Grenfell & James 1988; Robbins, 1993; Delanty, 2001), the intellectual field of university education is conceptualised as a field with a high degree of autonomy in that it generates its own values and behavioural imperatives which are relatively independent from forces emerging from the economic and political fields.

The field is structured in hierarchy in the sense that agents and institutions occupy dominant and sub-ordinate positions. These positions depend on the amount of specific resources that are possessed in relation to other occupants. Bourdieu refers to these field-specific resources as 'capital'. Capital may be viewed as the specific cultural or social (rather than economic) assets that are invested with value in the field which, when possessed, enables membership to the field.[48] The type of capital operating in the field of university education is an institutionalised form of cultural capital which has generally been termed 'academic' capital. In some instances (see *Homo Academicus*, 1998) Bourdieu distinguishes between two forms of capital: 'academic capital', which is linked to power over the instruments of reproduction of the university body; and 'intellectual' or 'scientific capital' which is linked to scientific authority or intellectual renown. In other instances however, for example in *State Nobility* (1996) the two definitions appear to merge and 'academic capital' is defined as an institutionalised form of cultural capital based on properties such as prior educational achievement, a 'disposition' to be academic (seen, for example, in manner of speech and writing), and specially designated competencies. It is in this second sense that 'academic capital' is used in this paper.

Bourdieu develops an understanding of the operation of practices occurring in higher education by the use of the concept 'strategy' which is understood as a specific orientation of practice. In contrast to the way the term is used in ordinary language, where it is based on conscious calculation, strategy in Bourdieu's theoretical apparatus is dependent on 'habitus', which as a result of socialisation, engenders in individuals a 'disposition' below the level of consciousness to act or think in certain ways; and on the network of objective relations between positions that agents or institutions occupy in the field. A major insight in Bourdieu's work is that even though the location of agents and institutions within a common field presupposes a minimum level of agreement around basic principles, the field of higher education is in fact not a product of total consensus but the product of a permanent conflict. The manifestations of these strategies are referred to as 'position takings' (Bourdieu, 1993, p. 35). These position-takings are inseparable from the objective positions occupied by the agent or the institution as a result of their possession of a determinate quantity of specific capital (Bourdieu, 1993). Agents and institutions individually or collectively implement strategies in order to

[48] Bourdieu defines capital in a formal sense as as accumulated labour in its 'materialised' or 'embodied' form which when 'appropriated' by agents 'enables them to appropriate social energy in the form of reified or living labour' (Bourdieu, 1986, p. 241).

improve or defend their positions in relation to other occupants. The importance of this theoretical framework allows Bourdieu to develop what he calls a 'general science of the economy of practices' within which university practices that purport to be 'disinterested' and hence non-economic, can be analysed as 'economic' practices in the sense that they are directed towards the maximising of symbolic gain (see, for example, Bourdieu, 1993). Bourdieu is adamant, however, that this does not mean reducing such struggles to the logic of economism.

Bourdieu's central concepts of 'field', 'capital', 'habitus' and 'strategy' are brought together in a relational framework to illustrate how it is precisely through this relative autonomy and through the hierarchical structure of the field that the social reproduction function of higher education is achieved. Higher education is conceptualised as a sorting machine which selects students according to an implicit social classification and reproduces the same students according to an explicit academic classification, which in reality is very similar to the implicit social classification (Bourdieu, 1996). While Bourdieu's use of a machine metaphor at first sight appears to confer a self-sufficient dynamic upon social structures (Bourdieu, 1996. p. 52), it is the internal cognitive workings that are of particular interest to him. Acts of cognition are implemented to select and consecrate what is classified as 'academic' and therefore what counts as valid criteria for entry and success in higher education. He labels these categories of perception 'academic taxonomies' (Bourdieu, 1996, pp. 17-19) and defines them as 'principles of vision and division' that structure academic judgements. Bourdieu's empirical studies (1979; 1988; 1996) reveal that this academic taxonomy is in fact organized according to the hierarchy of qualities commonly ascribed to the dominant group. The education system therefore designates those endowed with cultural capital, which is generally inherited as a result of social origin, as 'academically talented.' In this way, higher education establishes a close correspondence between the social classification at entry and the social classification at exit without explicitly recognising, and in most cases denying, the link between social properties dependent on social origin (such as class) and academic selection and evaluation.

Through the concept of homology Bourdieu emphasises the functioning of higher education as a relatively autonomous field which nevertheless contributes to social inequality. He defines homology as a 'resemblance within a difference' (Bourdieu & Wacquant, 1992, p. 106) and the concept appears to apply to both structural and functional qualities of fields. Each field mirrors the social space in having its own autonomous and heteronomous poles, its own dominant and dominated agents and institutions, its mechanisms for reproduction and its struggles for usurpation and exclusion. Many of the strategies function as 'double plays' (Bourdieu, 1996, p. 271) in the sense that although these field-specific strategies are not expressly conceived as such nor the product of conspiracy amongst the dominants in the various fields, they are objectively organized in such a way that they contribute to both the accumulation of field specific capital and to the reproduction of the structure of 'social space', defined by Bourdieu in an under-theorised way as 'class structure' (Bourdieu & Wacquant, 1992).

The higher education system thus acts as a 'relay' in that it reproduces the principles of social class and other forms of domination under the cloak of academic neutrality. It also acts as a 'screen' (Bourdieu, 1996, p. 36) which permits the realisation of social classification in guises that allow it to be accomplished invisibly. In this way, universities contribute to the 'misrecognition' and therefore 'naturalisation'[49] of structures of domination.

SOUTH AFRICAN UNIVERSITIES IN TRANSITION

In this section of the chapter Bourdieu's theoretical concepts are applied to the relationship between universities and forces for change in South Africa in the period 1985-1990 in order to illustrate their value (see Naidoo, 1999). The period of the study, which directly preceded the negotiated settlement that the Afrikaner Nationalist Government was compelled to enter into with revolutionary black organizations, has been characterised as a 'pre-revolutionary' one (Price, 1991). The dominance of the Afrikaner Nationalist government, which had segregated and hierarchically stratified South African society in line with the policy of apartheid, was challenged by a powerful mass-based political movement, the Mass Democratic Movement (MDM). The MDM, which had close links to the banned African National Congress (ANC), aimed to dismantle and replace the structures of 'racial capitalism' with alternative social, political and economic arrangements (Lodge and Nasson, 1991). In an attempt to stem the tide of revolt, the government initiated a series of minor reforms in 1985 which included the lifting of the 1959 legislation which segregated universities according to race.

In this period of political rupture between 1985 and 1990, alternative belief systems from the MDM drew on nationalist and socialist ideology to question the independence of universities from the political context. In a society which comprised 69 per cent Africans and 17 per cent whites, high status universities remained almost exclusively white. The de-segregation of higher education resulted in unremitting pressures on these universities to suspend highly selective meritocratic arrangements and to develop policies based on notions of disadvantage and redress, particularly in relation to African students, who had suffered extreme forms of discrimination under apartheid. Universities were also expected to transform themselves from elite to mass based institutions. The aim of the research was to analyse the translation of the socio-political forces within universities in the development of new admission policies. While the original study focussed on a range of forces emerging from the state and capital, this paper will focus exclusively on the forces arising from the MDM.

---

[49] Bourdieu explains this effect of 'naturalisation' by proposing that existing relations of domination are maintained because individuals' minds are structured according to cognitive structures that are issued out of the very structures of the world and thus perceive the organization and distribution of power in society as 'natural' (Bourdieu & Wacquant, 1992, p. 168). Education plays a key role in this process by converting and consecrating, and thus naturalising, social divisions in terms of both cognitive classifications and material privilege.

The South African Field of University Education

The first step of the research process was to construct the field of university education in South Africa. An analysis of the specific capital that was valorised in the field, the historical evolution of the relationship between the political context and the different types of universities, and their own position-takings, resulted in the field being hierarchically structured into a dominant, an intermediate and a subordinate tier. The dominant tier consisted of white English-medium universities which were set up in the colonial era for the British community. In 1959, under apartheid laws, these institutions were designated exclusively for white students. These universities were research intensive, internationally recognised and offered high-status postgraduate programmes. In the intermediate tier were the Afrikaans-medium universities, also designated white, which were set up by the Afrikaans community during the Anglo-Boer war. Universities which were set up for the different groups of black South Africans were in the subordinate tier. These institutions were poorly resourced, had almost no research infrastructure and offered low levels of study such as diplomas. The field of higher education was also structured into autonomous and heteronomous sectors according to the degree of state control and the extent to which institutions were able to reject external political determinants and obey the specific logic of the university field. The white English-medium universities were relatively free of direct state control and were positioned in the autonomous sector. They distanced themselves from the political context and although they opposed the segregation of universities, for much of their history they did not take a formal stance against other aspects of apartheid. They adhered to the internal laws of the field and developed products destined primarily for the intellectual field. The white Afrikaans-medium universities, located in the heteronomous sector, identified with the apartheid political project and contributed to the reproduction of apartheid culture (Davies, 1996). Black universities were also located in the heteronomous sector and were subject for much of their history to direct and violent state control and repression.

In this section, I will draw on material from two case study institutions located in contrasting positions in the field to illustrate how the particular historical development and the internal structure of the field mediated socio-political forces. Dominant representations rather than the micro-politics of each university will be presented. In order to capture the processes of change, Bourdieu's theoretical framework was supplemented with methodological strategies based on materialist approaches to discourse analysis. Interviews and institutional documents were therefore discursively analysed to capture the principles and underlying assumptions structuring accounts of admission policy development. This approach will be outlined in later sections of the chapter.

The first case study, which appears under the pseudonym Mt Pleasant University, was an elite historically-White English-medium university located in the dominant and autonomous sector of the field. The second case study, which appears under the pseudonym Freedom University, was a historically-black university which was designated for that sector of the black community designated

'Coloured'. Freedom University was located in the subordinate and heteronomous sector and was subject to direct state control. The two universities developed divergent and controversial admission policies in the 1985-1990 period.

Mount Pleasant University

At one level the political crisis appeared to have had an unprecedented impact on Mt Pleasant University's position of independence from the political field. For the first time since its inception, the University issued a public statement opposing apartheid and the Nationalist Government. However, there was less of a shift in relation to admission policies. In the face of increasing pressure for change, the traditional admission policy (hereafter referred to as the Standard policy) was maintained and strengthened. This policy allocated more than 90 per cent of student places and admitted those who achieved the highest scores in the school leaving examination (the matriculation examination). The vast majority of African students who achieved poor grades from racially segregated and inferior schools therefore continued to be excluded.

Documents and interviews especially at faculty level reveal a collective denial of social and political factors that might impact on universities, even in a country that appeared to be on the brink of a revolution. The institutionalised belief that entry to the field should be based purely on academic criteria and should be detached from social categories such as race, class or gender functioned to exclude moral and political pressures to address the issue of disadvantage. In addition, in the face of declining numbers of white applicants, a surge in black applications and an improvement in black matriculation results, the university implemented a number of additional policies. The argument was that these policies were necessary because the pool of talent in the country had decreased. A policy to limit undergraduate growth was instituted and the places available in a number of first year undergraduate courses were reduced. Finally, students for whom English was a second language (which included the majority of black students) were required to take an English language proficiency test. These policies acted as a further barrier to the admission of black students.

Forces for redress were instead deflected to an adjunct admission route, the Selective policy, which admitted a small number of African students. In the period of study, no more than one percent of places was allocated by the Selective policy and these student were generally admitted to sub-degree courses, thereby insulating mainstream programmes from students from non-traditional backgrounds. In one sense, the admission of students on the grounds of ethnicity eroded the strict operation of academic principles. However, the concept of 'academic potential' became part of the institutional discourse and students were admitted depending on their results in a test designed to measure academic potential. The implementation of 'latent' academic ability as a criteria for selection thus displaced the social criteria and re-inscribed academic principles, and indeed, academic merit.

The judgements implemented by powerful agents at Mt Pleasant University did not appear to be conscious and direct responses to external forces or, as some commentators have implied, inherent racism on the part of deans who attempted to exclude black students from an institution that was predominantly white. My analysis of interviews indicated that the systems of classification that were implemented appeared to be part of an orientation to conserve institutional arrangements which a new system of admissions could potentially threaten. New criteria were resisted because academics were concerned that an influx of students from poorer educational backgrounds would threaten institutional arrangements such as the proportion of time allocated to research and would endanger the high progression rate of the university. It was feared that these changes would endanger the status of the university in the intellectual field. Although informants vehemently denied any association, analysis of the legitimate academic capital (or what informants understood to be 'typical' students) were found to coincide closely with qualities possessed by students from the white schooling system. Thus although powerful agents such as deans did not appear to be excluding students in any direct sense on the grounds of ethnicity, their strategic interventions to conserve the traditional properties constituting admissions and to protect traditional academic practices functioned to exclude the majority of African applicants.

Freedom University

The second case study university, which appears under the pseudonym Freedom University, was located in the subordinate sector. Freedom University took the opportunity offered by the political crisis to challenge its status as an ethnic state-controlled university. The university forged an alliance with the MDM which resulted in the university remaining locked in the heteronomous sector. However, the political stance against apartheid and its aim of developing an alternative model of university education attracted a significant number of radical academics with high levels of academic capital. The University's position-taking and the influx of academic capital resulted in the university ascending to a dominant position relative to other black universities in the subordinate sector of the field.

Freedom University challenged the academic principles on which entry to the field was granted. The principle of academic merit, which was represented by dominant universities as fundamental in determining who would gain admission, was systematically eroded. Matriculation results as an indicator of academic merit was regarded as inappropriate in a society in which black students were forced into a poorly resourced inferior school system. A direct relationship between the operation of academic merit, the exclusion of black students and the system of apartheid was construed. Informants implied collusion between the elite white universities and the government in maintaining a racial hierarchy in South Africa.

Admissions strategies drew on discourses embodying the notion of 'redress' and 'rights' from the political field to construct new categories of admission and to argue that admission policies should compensate those who had been denied university access through apartheid policies. The university developed a non-selective and open admissions policy termed the 'Peoples Admission Policy' through which all students who met the statutory conditions for entry to higher education were accepted. In addition, in contrast to policy trends at other universities, Freedom University experienced increased student growth.

In 1989 the university acknowledged that the rapid growth rate had placed unbearable pressures on university staff and facilities. A new admission policy was developed, incorporating two mechanisms. The first mechanism, presented as a continuation of the open admission policy, consisted of computerised random selection through which eighty per cent of student places were allocated. The second consisted of the application of various criteria through which the remaining 20 per cent of student places were filled. Social criteria such as race, class and gender were applied. In addition, all students who showed 'academic achievement' were automatically admitted as part of the 20 per cent of accepted students.

At one level, the policy developed in 1989 drew directly on political discourses to replace the academic criteria dominant in the field with criteria based on principles of random and social selection. But at the same time academic principles were re- introduced by automatically selecting students who had achieved high matriculation grades. Analysis of the representations of the divergent classificatory principles underlying selection criteria reveals an almost complete absence of any recognised relationship of contradiction between the two policy discourses. For example, when discussing student selection through social criteria, the matriculation score was presented as an inappropriate measure. When discussing criteria based on academic skills, the matriculation score was presented as an appropriate indicator of success at university. This perhaps emphasises the fact that individuals are structured by the field while continuing to structure it. However, one consequence of the ambiguity of the admission policy was that it led to different groups of students claiming the 'right' to be prioritised for admission on the basis of either possessing good academic grades or possessing attributes of greater social disadvantage relative to other students. These challenges to admission policy which led to student strikes and other political action contributed to the crisis that engulfed the university in the 1990s.

A Mediating Field

The strategies developed by the two universities must be situated in the specific context of the 'pre-revolutionary' political situation during which mechanisms maintaining the relative autonomy of the field of university education from the socio-political context appeared to be partially and temporarily suspended. Alternative belief systems drawing on socialist principles became imbedded in popular consciousness and the principles structuring the field of university education were challenged. In this context, Mt Pleasant's position of dominance was threatened by the actions of a subordinate institution. The threat posed by Freedom University may be illustrated by the fact that each Dean at Mt Pleasant, and the majority of senior managers, with no prompting from the interviewer, launched an attack on the admission policies developed by Freedom University.

The findings also point to the importance of the historical evolution of relations between institutions and the socio-political complex. Such relations influenced the extent to which specialised structural and discursive mechanisms were developed. The relative independence of Mt Pleasant led to the development of complex discursive mechanisms independent of the political and economic context. These mechanisms were deployed to reposition, exclude and restructure political forces in the development of a differentiated but relatively coherent admission policy.

In contrast, Freedom University's close relationship to the socio-political complex led to an institutional context not conducive to the development of discursive mechanisms independent from external social and political discourses. The University did not appear to have recourse to specialised mechanisms that it could deploy in response to the contesting socio-political forces impacting on it. It appeared to respond to the political crisis by simply excluding or appropriating, rather than restructuring socio-political forces that emerged from the political field.

The findings above also point to the significance of the location of institutions in the internal power relations of the university field. Admission policy, which served to publicly codify the appropriate capital required to enter the field, became a crucial locus of struggle: it served to legitimise or delegitimise the principles underlying the recognition of existing capital in the field. Mt Pleasant's position of structural dominance oriented powerful agents in the institution to relate to external forces in such a way that the position of dominance was protected. Socio-political forces were excluded from the standard admission policy, and deflected to an adjunct policy. Further restructuring occurred in relation to the Selective policy to ensure that traditional academic principles underlying admissions were, in the final analysis, conserved. In this way Mt Pleasant University sought to ensure that the academic principles underlying the recognition of legitimate capital in the field was maintained and that its position of structural dominance was protected.

The development of admission policy at Freedom University can be seen as a strategic action taken by an institution located in the dominant fraction of the subordinate sector of the field during a time of social upheaval. External political principles were imported to erode the academic principles upon which power relations in the field were based and which placed the University in a subordinate position. However, the growing dissonance between an admissions policy based on social criteria and the internal logic of the field, as well as the University's own ascendance in the field, also resulted in the appropriation of discourses around academic merit which were valorised by dominant institutions. An overarching institutional discourse that could restructure such discourses in a coherent manner was unavailable, resulting in an ambiguous and contradictory admissions strategy.

The extent to which the two case study universities contributed indirectly to the reproduction of the racial hierarchy in South Africa is complex. At one level, a simple indicator of the extent to which such policies reproduced racial stratification may be measured by the proportion of students classified African, originating from the most dominated sectors of society, that were admitted to each university. In this sense, the admission policies of Mt Pleasant University, which admitted mainly white students from dominant positions in society, while largely excluding African students, can be seen to contribute to have reproduced fundamental racial divisions. The policies developed by Freedom University, on the other hand, which resulted in the largest enrolment of African students relative to all other non-African universities in the country, can be seen to have challenged racial divisions.

What is unclear, however, is the extent to which each university contributed to placing students in more privileged positions in society at the point of graduation. The admission strategies of Mt Pleasant University could have enabled an elite minority of African students to access more privileged positions in society. On the other hand, while Freedom University provided access to large numbers of African students, it is debatable whether they were placed in a more privileged position in the social structure of South Africa. Researchers have indicated for example that alternative educational strategies may inadvertently entrench the hierarchy between privileged and disadvantaged students in capitalist society by conferring on students a qualification that is not recognised as sufficiently 'academic' by other universities or by employers (see, for example King, 1993). In this sense, the admissions strategies developed by Freedom University, which were constructed by the direct appropriation of political principles, may not have resulted in a redistribution of academic capital to Black students; nor did such strategies lead to the development of an alternative type of capital which could erode the dominant principles of academic capital to function in a relatively autonomous field.

THE CONTRIBUTIONS OF FIELD THEORY

The South African case study, Bourdieu's research and research in other national contexts (Tomusk, 2000; Deer, 2003) indicate the contribution of the concept of 'field' to sociological understandings of the relationship between higher education and society. Research has tended to conceptualise universities as closed systems which are detached from the socio-political complex or as reflections of external power relations within a determinant social structure. More recently, a number of sociologists have placed the relationship between socio-political forces and universities firmly on the agenda. However, a conceptual apparatus through which the macro forces and institutional practices could be analytically related has not been adequately developed. Researchers have in general theorised two conceptual levels: the one level which sees policy as being determined at the level of macro-social relations and a second level which allows considerable explanatory autonomy to human agency within institutions. The relationship between the two often appears to be either largely arbitrary or alternatively, the macro forces appear to provide the theoretical backdrop while the analysis at the level of the institution provides descriptive evidence for the grand theoretical framework.

The incorporation of Bourdieu's framework, by conceptualising universities as existing in specialised fields, alerts researchers to the interaction of external forces with the logic and structure of the field of university education and illustrates how external determinants may be restructured, repelled or even reversed. Theoretical representations of universities as existing independently of one another are also replaced with an understanding of universities as imbedded in complex relations of power with other universities. University responses can therefore be conceptualised as a type of 'position-taking', which is dependent on the institutions structural position in the field and which is oriented to safeguarding or improving this position. Bourdieu's theoretical constructs therefore make a significant contribution to understanding first, the relationship between macro socio-political forces and universities and second, why institutions within the same system develop divergent strategic responses to external pressures.

Bourdieu's 'field' concept also has particular ramifications for the relationship between the massification and democratisation of higher education. Bourdieu's work illustrates that it is precisely because the field of higher education consists of institutions that are classified in hierarchy that it can fulfil this classificatory function. Through acts of subjective and objective selection, a high proportion of students originating from different regions in social space are selected by institutions which traditionally supply the very same regions of social space. The structuring of the field of higher education, whilst establishing internal homogeneity in terms of social origin and disposition within each institution, establishes stark differences in social origin and therefore disposition between the student and staff populations in institutions positioned at different levels of relative hierarchy. It is therefore not individual universities but higher education as a system that contributes to reproducing and legitimating the 'ensemble of distances' that constitute social structure (Bourdieu, 1996).

Limitations and Further Developments in the Work of Bourdieu

In this section, I will outline two important limitations related to Bourdieu's focus on social reproduction which are of relevance to the concept of 'field'.

First, Bourdieu's methodological strategies cannot be easily applied to contexts characterised by social conflict and change. In his empirical work, Bourdieu analyses the relationship between educational institutions and political forces by identifying the underlying principles structuring activities in universities and illustrating how these principles are related (and in fact reproduce) the dominant principles (mainly based on class) structuring society. This analytical strategy results in two limitations. First, the exclusive focus on the dominant principles structuring society excludes an analysis of social forces which are strong enough to challenge dominant forces but too weak to entirely displace such forces. Second, Bourdieu's mapping of the underlying principles structuring the activities in the field of university education exclusively onto the dominant principles structuring society may be seen to be analogous to viewing cinematic stills exclusively at the beginning and at the end of a sequence of actions - we are not able to view the series of steps by which the initial action relates to the final action. This methodological strategy renders the process by which educational or social principles are produced and reproduced invisible to analysis. By excluding the process by which such principles are produced, Bourdieu excludes the key mechanisms by which universities appropriate social forces to reproduce selected principles while excluding and displacing other social forces. His approach therefore leads to less illumination of university practices.

On the other hand, however, Bourdieu's thinking tools do not prevent the incorporation of other perspectives to complement the limitations. One way to ensure that process, contradiction and change are included in Bourdieu's framework is to incorporate materialist approaches to discourse analysis. Such language-based analyses have been adopted by a wide range of sociologists including Rizvi & Kemmis (1987) and Ball (1994). In the above-mentioned accounts, policy is conceptualised as emerging from contested relations of power within institutional and social settings and is therefore the product of negotiation, struggle and compromise. In this sense, policy texts and statements may be conceptualised as ideologically constructed products. Representations of policy will therefore embody, in implicit and explicit form, many of the contradictory and contesting social and political forces that were part of its production. The application of methods of discourse analysis to documents and interview transcripts is therefore likely to reveal the constitutive discourses that structure such texts and is therefore likely to render the processes of policy development more visible to analysis. The incorporation of methods of discourse analysis within the meta-theoretical framework of Pierre Bourdieu (see Naidoo, 1999) therefore makes it possible to analyse two sets of relations: the principles and underlying assumptions structuring various discourses in relation to other discourses and the relationship between the different discourses and the structural positions held in the institutional field within which such discourses are produced.

The second limitation in Bourdieu's general theory of fields is that there is little focus on the internal content of a given field. LiPuma (1993) notes that Bourdieu holds an 'absolute substantive theory of arbitrariness' to the extent that the content of a cultural product such as a text, a painting or a theoretical development is arbitrary and could just as well have been replaced by an alternative product to serve the same function. All that is necessary in Bourdieu's model is for the cultural product to produce distinction as a means of expressing and reproducing relations of class inequality. In relation to his work on higher education, Bourdieu focuses on structural positions and position takings in the struggle to impose and monopolise academic authority. The content and internal structuring of knowledge is therefore excluded from analysis and academic products are perceived as mere emanations of the respective positions occupied in the field which relate in a homologous pattern to social structure. This is a key gap in Bourdieu's work.

The work of Basil Berstein (1996) on the intrinsic structuring of knowledge indicates that differential knowledge structures have a decisive impact on the social relations and academic products in any given field. Moore & Maton (2001), drawing on Bernstein, show how developments in mathematics can be traced to the discipline of mathematics as a specific knowledge formation, rather than as a mere reflection of field relations. Bernstein's proposition that sociological accounts of the relationship between education and society need to incorporate both symbolic structures and social relations as part of one system for the purposes of analysis is therefore crucial to the further development of Bourdieu's framework.

CONCLUSION

One question is the extent to which Bourdieu's field concept remains relevant in the context of current developments in higher education. Bourdieu's theoretical and empirical work was developed in the context of a 'social compact' that evolved between higher education, the state and society over the last century and which led to the insulation of universities from direct market pressures (Slaughter & Leslie, 1997; Marginson & Considine, 2000). However, Government perceptions of higher education as a lucrative service that can be sold in the global marketplace has begun to eclipse the social and cultural underpinnings of the social compact. Relatedly, the belief that universities require a relative independence from political and corporate influence to function optimally, which was in turn linked to the need for guaranteed state funding, academic tenure and professional autonomy, has been eroded. These changes have resulted in the implementation of funding and regulatory frameworks which revolve around neo-liberal market mechanisms and new managerialist principles (Avis, 1996; Williams, 1997; Deem, 2001). Such developments have led to a growing and relatively unmediated influence of economic forces on the field of university education which has in turn resulted in the undermining of cultural capital and the valorisation of economic capital.

While Bourdieu has not focused on the growing commodification of university practices and products, his theoretical concepts remain valid. First, Bourdieu's framework is predicated on the knowledge that the relative autonomy of fields varies from one period to another and from one national tradition to another. The conditions for the relative autonomisation or de-autonomisation of fields has also been developed and clarified (Bourdieu, 1992; Vandenberghe, 1999). Second, his insight that the field of higher education is not a product of consensus but the product of a permanent conflict is important. In an era where both academic and economic forces exert powerful structuring effects on universities, the field is likely to become the locus of power struggles over the legitimate capital required to attain dominance. Preliminary work on the effects of forces for commodification (for example, Naidoo & Jamieson (2002)) indicates that elite universities are more likely to possess the financial and cultural resources to restructure or redeploy forces for commodification in order to re-legitimise academic capital and so protect their position of dominance in the field. Subordinate universities, which admit students from disadvantaged backgrounds, are more likely to be buffeted by market forces. Bourdieu's framework is therefore likely to continue to play an important role in contributing to sociological understandings of the extent to which commodification is likely to erode or exacerbate social equity.

REFERENCES

Avis, J. (1996). The enemy within: quality and managerialism in education. In: J. Avis, M. Bloomer, G. Esland, D. Gleeson, & P. Hodkinson (Eds.) *Knowledge and nationhood: Education,politics and work.* London & New York: Cassell.

Ball, S.J. (1994). *Education reform: A critical and post-structural approach.* Buckingham, Open University Press.

Bernstein, B. (1996). *Pedagogy, symbolic control and identity: Theory, research, critique*. London: Taylor & Francis.
Bourdieu, P. (1986). The forms of capital. In: G.R. Richardson (Ed.) *Handbook of theory and research for the sociology of education*. New York: Greenwood Press.
Bourdieu, P. (1988). *Homo Academicus*. Cambridge: Polity Press.
Bourdieu, P. (1992). *The rules of art: Genesis and structure of the literary field*. Cambridge: Polity.
Bourdieu, P. (1993). *The field of cultural production*. Cambridge: Polity Press.
Bourdieu, P. (1996). *The state nobility*. Cambridge: Polity Press.
Bourdieu, P. & Wacquant, L.J.D. (1992). *An invitation to reflexive sociology*. Cambridge: Polity Press.
Davies, J (1996). The state and the South African university system under apartheid. *Comparative Education*, 32(3), 319 –32.
Deem, R (2001). Globalization, new managerialism, academic capitalism and entrepreneuralism in universities: is the local dimension still important? *Comparative Education*, 37(1), 7-20.
Deer, C. (2003). Bourdieu on higher education: the meaning of the growing integration of educational systems and self reflective practice. *British Journal of Sociology of Education*, 24(2), 195-207.
Grenfell, M. & James, D. (Eds.) (1988) *Bourdieu and education: Acts of practical theory*. London: Falmer.
Endry, J. & Bunting, I. A. (1993) *Universities in South Africa*. Cape Town: University of Cape Town.
King, K. (1993). *Education policy in a climate of entitlement*. Unpublished discussion paper.
LiPuma, E. (1993). Culture and the concept of culture in a theory of practice. In Calhoun, C., LiPuma, E. & Postone, M. (Eds.) *Bourdieu: Critical perspectives*. Cambridge: Polity Press.
Lodge, T. & Nasson, B. (Eds.) (1991). *All, here, and now: Black politics in South Africa in the 1980s*. (London: Hurst and Co. and Ford Foundation.
Marginson, S. & Considine, M. (2000). *The enterprise university: Power, governance and reinvention in Australia*. Cambridge: Cambridge University Press.
Moore, R. & Maton, K. (2001). Founding the sociology of knowledge: Basil Bernstein, intellectual fields and the epistemic device. In Morain, A., Neves, I., Davies, B. & Daniels, H. (Eds.) *Towards a sociology of pedagogy: The contribution of Basil Bernstein to research*. New York: Peter Lang.
Naidoo, R. (1999). Admission policies and the politics of access: A case study of two South African universities (1985-1990). Unpublished PhD dissertation. Cambridge: University of Cambridge.
Naidoo, R & Jamieson, I (2002). *Consumerism in higher education and its implications for a high skills society*. Presented to the European Conference for Educational Research. Lisbon.
Price, R.M. (1991). *The apartheid state in crisis: Political transformation in South Africa 1975-1990*. Oxford: Oxford University Press.
Rizvi, F. & Kemmis, S. (1987). *Dilemmas of reform: The participation and equity program in Victorian schools*. Geelong: Deakin Institute for Studies in Education.
Robbins, D. (1993). The practical importance of Bourdieu's analyses of higher education. *Studies in Higher Education*, 18(2), 151-163.
Slaughter, S. & Leslie, L. (1997). *Academic capitalism: Politics, policies, and the entrepreneurial University*. Baltimore: Johns Hopkins University Press.
Tomusk, V. (2000. Reproduction of the 'state nobility' in Eastern Europe: Past patterns and new practices. *British Journal of Sociology of Education*, 21(2), 269-282.
Vandenberghe, F. (1999). 'The real is relational': An epistemological analysis of Pierre Bourdieu's generative structuralism, *Sociological Theory* 17(1), 32-67
William, G. (1997). The market route to mass higher education: British experience 1979-1996. *Higher Education Policy*, 10, 275 -289.

Rajani Naidoo
School of Management
University of Bath, Bath, United Kingdom

J. KIRSTEN TURNER & BRIAN PUSSER

12. ADVANCING LEADERSHIP AS A PUBLIC GOOD:

The Challenge of Postsecondary Selectivity in Virginia, USA

INTRODUCTION

Colleges and universities have long been charged with developing the qualities of leadership in students and in producing a significant cohort of future leaders for communities and the broader society. Leadership development has been seen as one of the primary public benefits of post-secondary education, and the importance of this benefit runs throughout the literature of higher education (Pascarella & Terenzini, 1991; Kerr, 2001; Ordorika, 2003; Bowen et al., 2005). Post-secondary institutions contribute to the development of community, state, and national leaders by providing education, networks, and relationships that help students engage in their communities, as well as on a state's larger political, professional, and social stages. These institutions prepare students for positions of leadership through various forms of education and socialization embedded in curricula, extra-curricular experiences, professional training, and credentialing (Bowen & Bok, 1998a; Astin & Astin, 2000). Through this process students may become role models within their home communities, first as college graduates and later as professionals and community leaders. In addition, a national cadre of community-based leaders is built over time, as students leave the institution and join communities throughout the state and beyond as successful professional and civic figures (Burkhardt & Zimmerman-Oster, 1999; Pusser, 2006a; 2006b).

While many sectors of post-secondary education contribute to the production of leaders and leadership, there is evidence to suggest that the most prestigious political, corporate, and community leadership positions often go to individuals who attend the most selective colleges and universities (Bowen & Bok, 1998a; Slaughter & Rhoades, 2004; Engell & Dangerfield, 2005). It has also been argued that selective institutions offer a different form of leadership development than non-selective institutions (Lemann, 1999), including greater increases in students' self-esteem and occupational aspirations when compared with attendance at less-selective institutions (Winter et al., 1981; Bowen & Bok, 1998a). The policies and rituals of selective institutions have also been shown to create a self-fulfilling prophecy in which students acquire a sense that they are part of an elite group and are society's future leaders (Spence, 1973; Colarelli et al., 1987). At the same time, Supreme Court justices, academic scholars, and institutional administrators alike have argued that a desirable form of post-secondary leadership development occurs at institutions that enrol highly talented students from diverse communities (Hurtado et al, 1999; Chang et al, 2003). When this occurs, two separate forms of leadership development are enhanced. First, institutions can help create leaders

Simon Marginson (ed.), Prospects of Higher Education, 229–250.

from a variety of communities, including communities traditionally underserved. Second, a diversified student body helps foster greater multicultural understanding, as it exposes the future political, professional, community, and social leaders to society's demographic and geographic sub-populations (Hurtado et al., 1999; Joint Statement of Constitutional Law Scholars, 2003; Chang et al., 2003).

However, despite the benefits a diverse student cohort brings to individuals, institutions, and communities, considerable research over the past two decades documents the under-representation in post-secondary education of students from a number of racial-ethnic groups, of various socioeconomic statuses, and from particular geographic locations (Bowen & Bok, 1998a; Advisory Committee on Student Financial Assistance, 2001; Hurtado & Wathington Cade, 2001; Advisory Committee on Student Financial Assistance, 2002; Astin & Oseguera, 2004; Bowen et al., 2005), particularly at selective higher education institutions, which tend disproportionately to enrol students from relatively homogeneous populations (Hearn, 1984; McDonough et al., 1996; Astin & Oseguera, 2004; Martin et al., 2003). Patterns of under-representation coupled with the inequitable distribution of access constitute a considerable challenge to selective public universities for which providing public benefits and developing a diverse set of leaders have long been fundamental to their missions (Casteen, 2001; Engell & Dangerfield, 2005).

Over the past decade significant efforts have been directed towards improving student diversity in selective public higher education institutions, notably through outreach programs, financial subsidies, and affirmative action programs. As efforts to increase equity and diversity in selective public institutions have come under increased legal and legislative challenge, states and institutions have implemented additional methods for distributing access (Olivas, 1999; Pusser, 2004). Among the most prominent programs are the 'percent plans' in Texas, California and Florida (Horn and Flores, 2003; Martin et al, 2003). Percent plans, which guarantee admission to a portion of the graduating class from each high school in a state, are underpinned by the notion of proportional representation, which suggests that the proportion of students from different backgrounds in an admission cohort should be similar to the distribution in the population of the state or nation (Wilds, 2000).

While there is considerable debate over how effective percent plans have been in increasing under-represented enrolments to date (Hebel, 2003; Horn and Flores, 2003; Marin and Lee, 2003), they have turned attention to the need for more detailed research on proportional representation and the distribution of access to selective public institutions, particularly in terms of its effect on the production of leaders within a state's geographic and demographic communities. We address the issue here using a longitudinal data analysis from one of the most selective public institutions in the United States, the University of Virginia.

The University of Virginia has an historical mission to create leaders at the state and local levels. That mission has been renewed through a contemporary push for greater access and diversity, while at the same time the institution has strongly increased its selectivity. The resulting effort to satisfy these inherently conflicting goals has been instrumental in shaping student admission cohorts, and thus, the distribution of access to a range of public benefits, including leadership development and community building, that the University endeavours to provide. While this research focuses on one elite public university in the United States, the distribution of the public benefit of leadership development and the effort to redress historical patterns of enrolment in elite public postsecondary institutions is a challenge shared by institutional leaders and policy makers across the world.

THE CHALLENGE

Over the last hundred years American higher education has been shaped by an increasing demand for access. Fuelled by such factors as the expansion of compulsory K-12 education, the Servicemen's Readjustment Act of 1944 (the GI Bill), the creation of federal and state financial aid, the increased admission of people of colour, women, and adult students, and the expansion of the community college system (Rudolph, 1990; Pascarella & Terenzini, 1991; Cohen & Brawer, 1996; Kane, 1999), the demand for admission to highly ranked institutions has far outstripped the supply (Kirp, 2003; Geiger, 2004; Bowen et al., 2005).

Increased demand led to the construction of criteria for selecting which students institutions would serve, and the admission requirements implemented at these institutions. The most selective institutions attracted and admitted the students they deemed most eligible. Institutions were increasingly ranked in a prestige hierarchy based in no small measure on their selectivity (Kirp, 2003; Geiger, 2004). One result of the linkage of selectivity and prestige has been the creation of a set of admission criteria that contribute to the construction of a meritocracy for students and the construction and perpetuation of a prestige hierarchy of institutions (Marginson, 1997; Lemann, 1999; Geiger, 2004; Bowen et al., 2005).

Only a few hundred institutions have admissions criteria that move beyond threshold eligibility requirements to incorporate competitive standards. Probably only a third of those could be characterized as 'highly competitive'. Yet it is in those highly selective institutions that the prominent contemporary contests over access and equity occur. It has been argued that over time the benefits offered by selective higher education institutions, particularly in the form of positional goods, make access to selective public higher education increasingly valuable, and in turn increasingly desirable. Winston (1999) has argued that due to their considerable financial resources the most selective institutions spend the most on education and – along with their high tuition prices - offer the highest per student subsidies. In essence, those who gain access to selective higher education institutions receive the highest institutional investment and positional benefits (McMahon, 1999).

While considerable efforts are being made to diversify selective higher education institutions on a number of dimensions, those institutions still tend to enrol students from a relatively homogeneous population (Martin et al, 2003; Astin & Oseguera, 2004; Bowen et al, 2005). These are often white students from families of high socioeconomic status, who are graduates of highly effective K-12 systems (Hearn, 1984; Tierney, 1984; Teranishi, Allen & Solorzano, 2004; Martin et al, 2003; Astin & Oseguera, 2004). A number of factors shape these enrolment patterns. Thomas (1979) and others have argued that certain demographic characteristics - including social class, race, and gender - have a strong correlation with college choice (Karen, 1988; McDonough, 1997; McDonough et al., 2002). This view is rooted in the Bourdieuian model of educational institutions as social institutions that transmit cultural capital, and the notion that students with high levels of cultural capital – cultural norms, values, networks, and behaviours that are particular to each social class – have an inherent advantage in the college access process. This research further suggests that the amount of cultural capital a student possesses at the time of college admission determines to a large degree which institutions the student applies to and where she or he enrols. Other researchers linking socioeconomic status to primary and secondary schooling have also suggested that admitting prospective postsecondary students on existing criteria privileges students who reside in school districts with significant resources, college preparatory orientation, and highly qualified instructors (McDonough, 1997; Teranishi & Parker, 2003; Kirst & Venezia, 2004; Bowen et al, 2005).

Over the past two decades many public institutions have become increasingly selective, as the number of applications has increased and standardized test scores, grade point averages, and class rank have begun to play a more important role in public institutions' admission decisions (Breland et al., 1986; Miller et al., 1991; Lemann, 1999; Wightman, 2003; Pusser, 2004). In addition, Pitter & LeMon (1991) found that disciplines in public institutions seen as offering better returns in the labour market are also becoming increasingly selective.

A number of scholars have begun to question the role of selective admission policies at public institutions in generating and preserving the public benefits of higher education (Giroux & Giroux, 2004). These scholars have argued that the criteria on which institutions base their admission decisions create barriers for low-income and disadvantaged youth who lack the cultural capital or the educational preparation to successfully apply, be admitted, and enrol in selective higher education institutions. It has also been argued that the predominant criteria - high-school grade point averages and standardized tests - are not reliable predictors of students' future academic performance, particularly for students from disadvantaged populations (Astin, 1990; Greenleaf, 1990; Wightman, 2003). According to Astin (1990), selective admission policies hinder those students who possess the academic abilities but lack the cultural capital to apply to college.

While becoming more selective, elite public and private institutions in the United States have also turned attention to increasing student diversity. Scholars have linked a number of public and private benefits to increasing the diversity of postsecondary institutions, including a levelling of the economic and social disparities that exist between different social classes, races, ethnicities, and genders and a broadening of the contribution institutions make to social equality (Bowen & Bok, 1998a; McMahon, 1999; Chang et al., 2003; Bowen et al., 2005). Bowen (1978) argued that higher education has 'an advantage over other avenues towards equality - such as graduated taxes and public assistance - because it can reduce the inequality of what people are and what they can contribute, not merely of what they get' (p. 12). Over the last 30 years, selective higher education institutions have attempted to increase access, notably through outreach, affirmative action and student subsidy programs. Institutions established such programs to help facilitate the entrance of students from under-represented populations into higher education (Bowen & Bok, 1998b; Pusser, 2001; Wightman, 2003).

However, due in part to post-secondary education's role as a generator of exceptional private benefits for graduates, in recent years affirmative action programs have come under increased political and legal challenge (Horn & Flores, 2003; Pusser, 2004). Given the challenges, some state systems have attempted to develop alternative admissions criteria and practices in order to continue the effort to create a more equitable distribution of access to selective higher education.

Proportional Representation and Education for Leadership

Proportional representation is a central concept in shaping admissions practices and procedures in post-secondary education. It endeavours to create a distribution of students within an institution resembling the distribution in the broader population (Wilds, 2000; Horn & Flores, 2003). Foremost have been 'percent plans' which attempt to distribute access to selective public institutions more equitably. California, Florida, and Texas have implemented such plans over the past decade. Percent plans admit a certain number of students from each high school into the selective public higher education state system, generally on the basis of high-school class ranking. The Minorities in Higher Education Report 1999-2000 (Wilds, 2000) stated that, compared to other diversity-related initiatives, enrolment based on proportional representation achieves the greatest impact in terms of diversity awareness and education. Richardson & Skinner (1991) argued: 'Groups traditionally underrepresented in higher education benefit the most from policies based on distributive justice [proportional representation]' (p. 23). The National Task Force for Minority Achievement in Higher Education recommended that every state system use proportional representation (Hurtado et al., 1999). However, the early percent plans appear to be only somewhat effective in diversifying the most selective public campuses where they have been implemented. Many students admitted to the colleges in question would have been eligible without the plans, although it does appear that the plans bring greater visibility to college opportunities, and may enhance the effectiveness of outreach programs that contribute to diversifying admissions cohorts.

The tension between broad access and selectivity has implications for leadership development. According to Milem & Hakuta (2000), 'as our society has become increasingly diverse, the need for leaders who represent diverse communities has increased dramatically' (p. 8). Society needs leaders from a variety of demographic communities, and particularly from communities traditionally underserved (Chang et al., 2003). Society also needs a cadre of political, professional, and social leaders who have been exposed to different populations and have developed a strong multicultural understanding. According to Nickens et al (1994), culturally competent leaders endeavour to understand the nuances of specific sub-populations and can lead in culturally intuitive ways. Diversifying the cadre of leaders educated in selective institutions is particularly important to enhancing the public good of leadership development (Nickens et al., 1994; Bowen et al., 2005).

Research on selective institutions has shown that students from under-represented populations were more likely to return to and serve in marginalized communities (Bowen & Bok, 1998b). As one example, African-American men who attended selective institutions were more likely than white men to hold leadership roles in civic organizations, especially those organizations that centred on social services, youth, and school-related activities (Milem & Hakuta, 2000). Similarly, African-American women who attended selective institutions were more likely than white women to hold leadership roles in community, social service, alumnae, religious, and professional groups (Bowen &d Bok, 1998a). When professional leadership was examined, similar results were found (Keith et al., 1985; Moy & Bartman, 1995; Komaromy et al., 1997). Black and Hispanic medical professionals were more likely than white medical professionals to concentrate on specialties affecting the medically underserved, and more likely to locate their practices in underserved areas (Wilds, 2000).

Diversity in the student body also helps to produce leaders better prepared to lead a multicultural and pluralistic society. Milem & Hakuta (2000) wrote that 'students who have been exposed to greater diversity are more likely to demonstrate increases in racial understanding, cultural awareness and appreciation, engagement with social and political issues, and openness to diversity and challenge' (p. 62). In short, to preserve legitimacy, the diversity within society must be reflected in the leadership of society. In order for postsecondary institutions to produce the greatest degree of leadership development and to preserve their legitimacy as social and political institutions, these institutions must endeavour to enrol a student body that reflects the broader society.

THE CASE OF THE UNIVERSITY OF VIRGINIA

As a selective flagship institution, the University of Virginia (UVA) develops many of the State of Virginia's political, professional, and community leaders. The University was founded by the third President of the United States, has graduated a multitude of national political leaders including US senators, US congresspersons, and federal judges, numerous state governors, representatives and judges, along with countless other public servants, community and business leaders. In addition, the University, through its outreach efforts, institutional publications and its administration, evidences a strong commitment to training and educating Virginia's future leaders. In a statement entitled *Preparing a generation of leaders for Virginia 2020*, UVA President John Casteen (2001) wrote: '[the University is] engaged today in educating the women and men who will begin to take leadership positions in society' (The 2000-2001 President's Report, p. 16). This contemporary commitment is based on a unique, historical mission to develop the state's leaders. The founder of the University, Thomas Jefferson, argued that colleges and universities should train leaders for the benefit of the society. Jefferson's vision was to educate the most academically talented citizens of the state, regardless of rank in society, in order to produce the strongest citizen-professionals and political leaders (Jefferson, 1818). Today, this commitment has been extended beyond social rank to encompass race, ethnicity, gender, and geographic location. Although originally a school of white male students, over the past 40 years the University has become increasingly diverse. In the 1950s the University became racially integrated, and in 1970 it became co-educational. Today the University of Virginia leads all U.S. public institutions in graduation rates for African-American students, 86 per cent within six years of enrolment. The national average for flagship institutions is 43.8 per cent (Special report: African-American college graduation rates: Intolerably low, and not catching up with whites, 2002).

The University of Virginia is one of the nation's most prestigious public post-secondary institutions. Since 1993 UVA has continually ranked near the top of all public institutions and among the top 25 for all institutions in various college rankings. For the past four years UVA has been ranked either first or second among public research universities in the *US News and World Report* annual rankings. As UVA has become more highly ranked it has also become more selective. In fall 2001 the University received over 14,700 applications for the 2980 available first-year spaces. Approximately 38 per cent of applicants received an offer of admission. The incoming class of 2001 had a mean combined SAT score of 1313, with more than 80 per cent of admits ranked in the top 10 per cent of their high school class. Given the competition among applicants for admission, it is unclear whether the required preparation and eligibility characteristics necessary for admission are equally distributed across the state and within its various sub-populations. It is also unclear how the distribution of eligibility characteristics shapes the distribution of access to the University of Virginia.

Methodology

To determine the distribution of student access to UVA three separate analyses were performed. First, demographic data from in-state, first-year, first-time, undergraduate student admission files were obtained for two years, 1992 and 2002. Students' race, ethnicity, gender, and zip codes were collected. Three different demographic profiles of in-state, first-year, first-time, undergraduate students were created for each year, one for applicants, one for admits and one for enrolees.

Second, a demographic portrait of the population of each zip code in the state was compiled. The 1990 and 2000 US Census data provided a baseline of the number of citizens (17-21 years of age) from each home city or county and zip code, as well as the number of 17-21 year-olds of a particular race, ethnicity, and gender in each zip code. Zip codes were used as a measure that allowed for a considerably more fine-grained examination of the data than cities, counties or high school attended would have afforded. There are 918 zip codes in Virginia

Third, to determine proportional representation, two proportionality indices were created. Each index represented a ratio. The first index-ratio represented the number of students in the admission cohort at the University of Virginia as a proportion of the overall 17-21 year-old population of the State of Virginia. It was determined by dividing the total number of students in the admission cohort at the University of Virginia by the total number of 17-21 year-old citizens in the State.

The second index-ratio was determined by dividing the actual number of 17-21 year-old citizens residing in each zip code by the first index-ratio. The resulting number predicts how many 17-21 year-old citizens from each zip code would be a part of the admission cohort if admission slots were distributed randomly.

Proportionality indices were also created for the socio-demographic sub-populations of both admissions cohorts. To do this, the total number of 17-21 year-old citizens from each socio-demographic sub-population in the state was determined when compared to the overall 17-21 year-old state population. These demographic statistics were used to determine a predicted value for each socio-demographic population in each admission cohort.

Randomness, Proportionality and Prediction

At the centre of the research is the question of whether the distribution of student access to one state flagship university is proportional to the demographic distribution of 17-21 year old individuals in the state. In designing this research the authors contemplated a variety of questions about whether that comparison is appropriate. Since preparation for college is closely linked to scarce resources, it may seem unlikely that student postsecondary access could be proportionately distributed. It has been suggested elsewhere that not all young people aspire to attend state flagships, or attend college. These are reasonable questions that require further clarification. Similarly, while we have constructed a model that assumes all 17-21 year-olds should be considered potential matriculants at the state flagship university, it would be useful to compare the demographics of the University's admit cohort to the population of 17-21 year-old high school graduates in Virginia, or the cohort of Virginia's 17-21 year-olds who have graduated from high school and taken the SAT.

In 2003, the United States Supreme Court held that there is a compelling public interest in institutions of higher education striving to build diverse student cohorts (Joint Statement of Constitutional Law Scholars, 2003); there is an educational benefit generated through increasing and diversifying access. The essence of proportionality is comparison between a random and an actual distribution of applications, admissions and acceptances of offers of admissions to a university. Our goal is to document what the distribution of access actually has been, and to further document how it is distributed within sub-populations of the State of Virginia, and we suggest that proportional representation is essential to preserving the legitimacy and viability of public institutions at all levels of education.

FINDINGS

The results from this study demonstrate an increased concentration of admissions over time in a relatively few suburban areas of the state. The degree to which some of these areas were over-represented was considerable.

Table 12.1. Top ten zip codes for admission cohort, 1992, Virginia, USA

most applicants	*n*	*most admits*	*n*	*most enrolees*	*n*
22015 Fairfax	124	22015 Fairfax	66	22003 Fairfax	45
22003 Fairfax	103	22003 Fairfax	65	22901 Charlottesville	43
22901 Charlottesville	94	22901 Charlottesville	60	22015 Fairfax	40
22032 Fairfax	90	22032 Fairfax	54	22101 Fairfax	39
22101 Fairfax	90	22101 Fairfax	54	22032 Fairfax	38
22091 Fairfax	79	22091 Fairfax	45	23113 Midlothian	31
23229 Henrico	72	23113 Midlothian	42	23229 Henrico	26
22153 Fairfax	69	22192 Prince William	39	22153 Fairfax	24
22192 Prince William	68	23229 Henrico	39	22030 Fairfax	23
23233 Henrico	65	23233 Henrico	36	22152 Fairfax	22
23464 Virginia Beach	65			22182 Fairfax	22
				22192 Prince William	22
				23233 Henrico	22
total	854		500		331

Many of the state's urban and rural communities sent significantly fewer students to the University than a random distribution would have produced. The degree to which certain areas of the state were above the random predicted values for applications, enrolments and admissions and other areas of the state were below the predicted value points to a significant disparity in the distribution of access to the University, and access to leadership development and the institution's positional goods and benefits. In sum, a significant majority of the admission cohorts were concentrated in a relatively small number of zip codes. The results also document an under-representation of African-American and Hispanic students in University of Virginia admission cohorts that has persisted over the period of study.

Table 12.2. Top ten zip codes for admission cohort, 2002, Virginia, USA

most applicants	*n*	*most admits*	*n*	*most enrolees*	*n*
22101 Fairfax	131	22101 Fairfax	76	22101 Fairfax	52
22015 Fairfax	113	22015 Fairfax	61	22015 Fairfax	40
23233 Henrico	108	22901 Charlottesville	61	20171 Fairfax	39
22901 Charlottesville	95	20171 Fairfax	49	22039 Fairfax	37
20171 Fairfax	90	22039 Fairfax	49	22901 Charlottesville	37
22192 Prince William	87	22003 Fairfax	48	22003 Fairfax	34
22003 Fairfax	82	23233 Henrico	48	23233 Henrico	34
22039 Fairfax	82	22032 Fairfax	42	23113 Chesterfield	30
23113 Chesterfield	69	22153 Fairfax	41	22153 Fairfax	29
22153 Fairfax	67	22192 Prince William	40	22032 Fairfax	28
22182 Fairfax	67			22192 Prince William	28
total	924		515		360

The Distribution

Students from eight zip codes made up 16 per cent of the in-state students in the 1992 enrolled class, despite the fact that these eight zip codes represented 3.9 per cent of the 17-21 year-old population (Table 12.1). These zip codes were significantly over-represented. The combined number of students they were predicted to enrol was 67 students, the actual number enrolled was 275 students.

In the 2002 admission cohort, 60.8 per cent of the state's zip codes were represented in students' applications; 48.2 per cent of the zip codes in the admitted class; and 41.2 per cent in the enrolled class (Table 12.2). Just 10 zip codes, or 1.1 per cent, produced 16.1 per cent of the applications, 17.1 per cent of the admits, and 17.9 per cent of the enrolees. While these zip codes were located in some of the most populated areas of the state, they were still far more prominent than predicted. Based on their 17-21 year-old population these zip codes were predicted to enrol a total of 73 students, but they collectively enrolled 330 students.

In contrast to many suburban areas, rural areas were collectively less prominent in the admission process than predicted. In 1992, the state's most rural zip codes were predicted to send 131 applications, yet sent only 100 applications. These zip codes were also well below predicted values for admits and enrolees (Table 12.3).

Table 12.3. The predictive distribution of sparsely populated zip codes, 1992

	applicants	*admits*	*enrolees*
Degree of proportionality	-1.3	-1.4	-1.5
Predicted value	131.20	156.86	168.24
Actual value	100	112	116

Table 12.4. The predictive distribution of sparsely populated zip codes, 2002

	applicants	*admits*	*enrolees*
Degree of proportionality	1.03	-1.1	-1.2
Predicted value	133.23	161.84	166.25
Actual value	137	144	143

This dynamic changed slightly in 2002, when there were greater numbers of applications than predicted from the sparsely populated zip codes. The number of rural admits and enrolees continued below the predicted values. This suggests that in 2002 rural students applied to the University in numbers greater than their 17-21 year-old population would have predicted but that did not result in substantially increased admittance or enrolment. In both the 1992 and 2002 admitted and enrolled classes, students from the state's rural areas did not access the University at levels proportional to their share of the state's 17-21 year-olds (Table 12.4).

Geospatial Mapping

Geospatial mapping of the 1992 and 2002 admission cohorts was employed to present these data for visual analysis. Select zip codes in northern Virginia (the western suburbs of Washington, D.C.), as well as zip codes in many of the state's other urban centres, such as Richmond and Charlottesville, reflect greater-than-predicted values. However, closer examination of urban areas presented a mixed picture. The geospatial maps indicate that pockets of certain urban areas were under their predicted values for application, enrolment and admission. For example, in the 1992 applications data, zip codes located in the southern and southeastern sides of Richmond were under their predicted values. The Tidewater region, one of the most populated areas in the state, was also largely under the predicted values. In 2002, the state's urban centres were predominantly over their predicted values (with the exception of the Tidewater region). In contrast, the rural communities of southwestern Virginia, for the most part, remained under their predicted values. There were changes in the overall distribution from 1992 to 2002.

Map 1
Proportionality of Zip Codes for 1992 Applicants

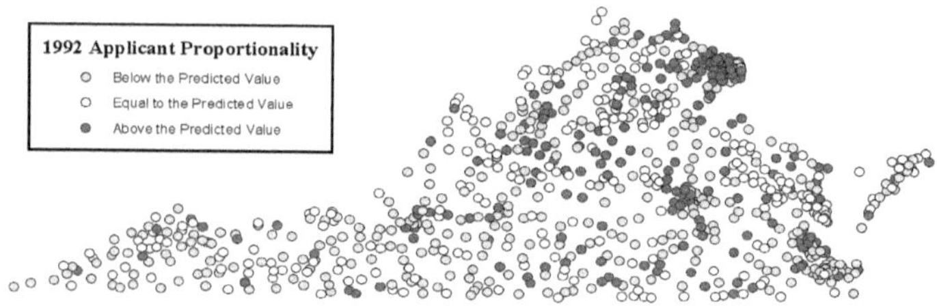

Map 2
Proportionality of Zip Codes for 1992 Admits

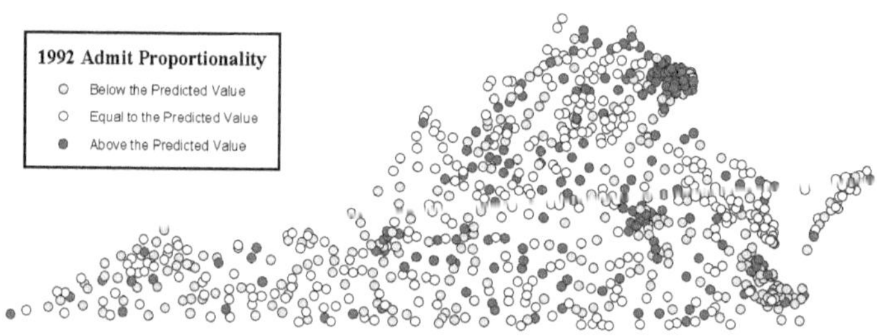

Map 3
Proportionality of Zip Codes for 1992 Enrollees

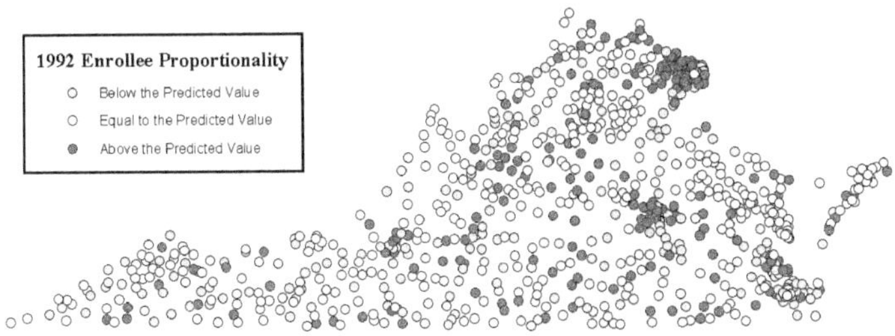

Map 4
Proportionality of Zip Codes for 2002 Applicants

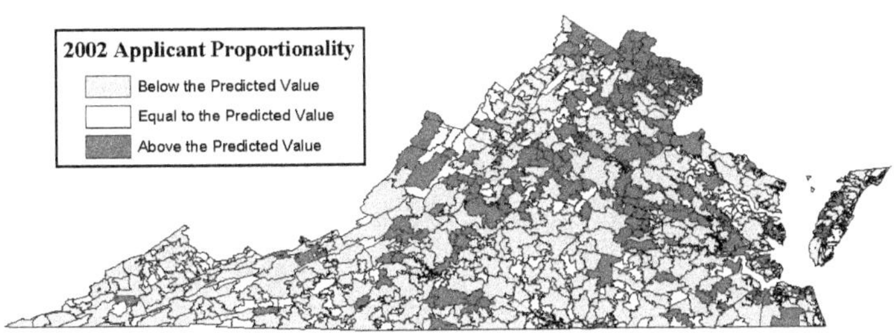

Map 5
Proportionality of Zip Codes for 2002
Admits

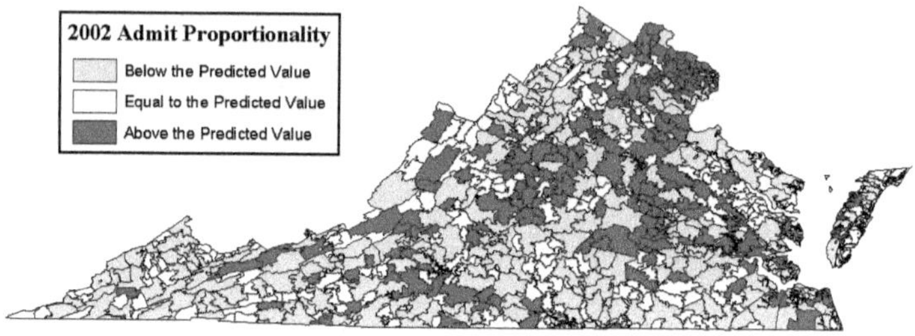

Map 6
Proportionality of Zip Codes for 2002
Enrollees

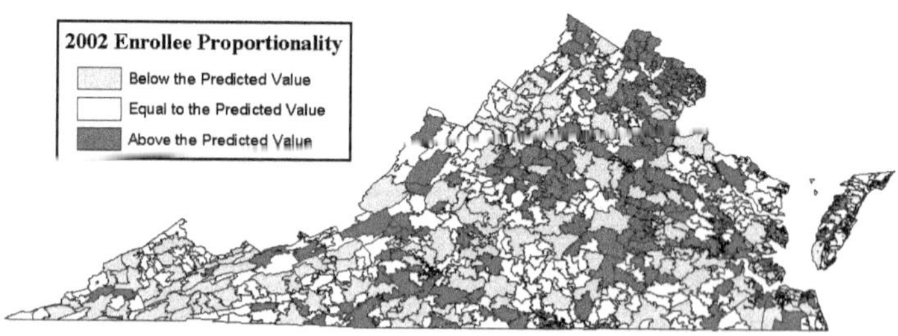

Table 12.5. Representation of the 1992 enrolees by demographic sub-population

Race and ethnicity	*percent of 1992 enrolee class*	*percent of 1990 17-21 year-old population*	*percent over- or under-predicted*
African American			
men	3.0	10.3	-70.9
women	6.8	10.2	-32.7
total	9.8	20.5	-52.2
Asian American			
men	5.6	1.6	+250.0
women	7.2	1.3	+453.8
total	12.8	2.9	+341.4
Hispanic American			
men	0.64	1.9	-66.3
women	1.33	1.3	+2.3
total	1.97	3.2	-38.4
White American			
men	33.55	37.9	-11.6
women	39.89	34.0	+15.6
total	72.84	71.9	+1.25
Other			
men	1.1	1.0	+10.0
women	1.5	0.5	+200.0
total	2.6	1.5	+73.3
men	43.9	52.6	-16.5
women	56.1	47.4	+18.4

Place Matters

Collectively, these data suggest that students who lived in certain geographic locations within the state were far more likely to apply, be enrolled and matriculate at the flagship university than were students from other locations. Based on the analysis of sub-populations in the data, and in keeping with the literature on college access, it appears that these disparities are likely grounded in disparities in family income, parent's level of education and local resources for elementary-secondary education. Such disparities in representation between the state's communities are particularly problematic in terms of distributing access and the resulting leadership development, positional goods and benefits state-wide. Enrolling a high number of students from a relatively small number of communities constrains the geographic and demographic distribution of post-secondary goods, and decreases the number of communities in the state that will benefit from flagship-trained students. It also limits the educational benefits generated for all students by a geographically and demographically diverse student body. To assure that the leadership development produced by the state's flagship is equitably distributed throughout the State of Virginia, access to that institution needs to be available to students throughout the state.

Table 12.6. Representation of the 2002 enrolees by demographic sub-population

Race and ethnicity	*percent of 2002 enrolee class*	*percent of 2000 17-21 year-old population*	*percent over- or under-predicted*
African American			
men	3.04	11.0	-72.4
women	5.43	11.0	-50.6
total	8.47	22.0	-61.5
Asian American			
men	5.9	1.9	+210.5
women	7.6	1.8	+322.2
total	13.5	3.7	+264.9
Hispanic American			
men	1.7	3.9	-56.4
women	2.0	2.7	-25.9
total	3.7	6.6	-43.9
White American			
men	29.42	33.3	-11.7
women	37.44	30.9	+21.2
total	66.86	64.2	+4.14
Other			
men	3.5	2.1	+66.7
women	4.0	1.4	+185.7
total	7.5	3.5	+114.3
men	43.5	52.1	-16.5
women	56.5	47.9	+18.0

Given that scholars have argued the most productive way to achieve a demographically diverse student body is through proportional representation (Richardson & Skinner, 1991; Hurtado et al., 1999; Wilds, 2000), the data from this research raise significant challenges.

Sub-Populations

Various demographic sub-populations in the 1992 and 2002 admission cohorts were not enrolled in numbers equivalent to their representation in the state's 17-21 year-old population. White-American students were enrolled in numbers quite similar to their presence in the 17-21 year-old cohort, Asian-American students and female students were present in numbers proportionally higher than in the 17-21 cohort, and African-American students, Hispanic-American students, and male students were present in numbers proportionally lower than in the 17-21 year-old cohort. Collectively, the data point to a distribution of access to the University throughout the state's demographic sub-populations that is quite different from the distribution within the state's 17-21 year-old population (see Tables 12.5 and 12.6).

Income, Post-secondary Attainment and Access

Research has linked a number of conditions and characteristics to success in the college admission process, such as familial socioeconomic status, parental levels of education, possession of high levels of cultural capital, and attendance at a high school with significant numbers of students preparing for post-secondary attendance (Karen, 1988; McDonough et al., 1996; McDonough, 1997; Bowen et al., 2005). By employing US Census data, two of these characteristics - post-secondary attainment and family income levels - were explored with respect to the 1992 and 2002 admission cohorts at UVA. This analysis suggests that 17-21 year-olds from areas of greater wealth and higher levels of educational attainment were far more likely to be represented in the University's admission cohorts.

In 1992, eight zip codes dominated the admission cohort at UVA. Collectively, the citizens of these zip codes were also wealthier than the rest of the state. Their average median family income was $63,147, which compared to state's median family income of $38,213. The citizens of these zip codes also reported high levels of post-secondary attainment; on average 51.4 per cent of the citizens (over the age of 25) living in these eight zip codes possessed a postsecondary degree. The state average was 24.5 per cent (American Fact Finder, 1990). The nine zip codes that dominated the 2002 admission cohort were also wealthier and more educated than other parts of the state. In 2000, the average median family income of these zip codes was $99,623. In comparison, the state's median family income was $54,169. The median price asked for a housing unit for these zip codes was $268,722 as compared to the state's median asking price of $87,600. One zip code, 22101 - in Fairfax, Virginia, which produced the most applicants, admits, and enrolees for UVA's 2002 admission cohort - had a median asking price for a housing unit of $863,600. The citizens living in these nine zip codes were also well educated. The percent of citizens over the age of 25 with a post-secondary degree was 59.0 per cent. The state's average was 29.5 per cent (American Fact Finder, 2000).

The Convergence of Place and Demographics

The concurrent geographic and demographic disparities suggest that certain demographic sub-populations could be concurrently above and below predicted values. In some zip codes Hispanic-American students in northern Virginia were under-represented relative to their presence in the 17-21 year-old cohort as Hispanic-Americans, but overrepresented in the 17-21 year-old cohort as residents of northern Virginia. In a number of zip codes Asian-American students in the Tidewater region were under-represented as residents of that region relative to their presence in the 17-21 year-old cohort, but over-represented relative to their presence in the 17-21 year-old cohort as Asian-Americans. These results hold peculiar implications for efforts to equitably distribute access to the University and thus to the institution's leadership development opportunities, positional goods and public benefits. Attempting to increase the representation of a demographic sub-population, such as African-Americans, may be most easily facilitated in the northern Virginia region; however, by doing so the existing concentration of enrolments from that region would be increased.

Furthermore, when the data are analysed according to zip code, such discrepancies become even more complicated. For example, in 1992, and to a greater extent in 2002, African-American students consistently applied, were admitted and enrolled in numbers above predicted values in northern Virginia, but consistently below predicted values in the southernmost zip codes of northern Virginia. Such results point to two different levels of access to the state's flagship institution for relatively similar demographic sub-populations (both sets of students were northern Virginian African-Americans). A similar pattern for African-American students was witnessed in the city of Richmond (the state capital) between the western suburbs and southeastern suburbs. These results demonstrate that understanding access at finer levels of geographic demarcation helps to more precisely account for patterns of the distribution of access.

These data present multiple challenges for efforts to proportionally distribute access. For example, to achieve African-American representation at the University reflecting both the geographic and demographic distribution of the state's 17-21 year-old population, it may not suffice to enrol greater numbers of African-American applicants from northern Virginia; it may also be necessary to enrol African-American applicants from more specific locations in those communities.

CONCLUSIONS

The analysis of this research presents a persistent and disproportionate pattern of access to a state flagship institution. While this study focuses on one highly visible and influential institution, it also points to the utility of case studies of the distribution of access at state flagships and other public institutions throughout the post-secondary system. The findings from this case study point to the difficulty in creating a flagship admission cohort that is both selective and proportionally representative of a state's demographic geography. The persistence of patterns of disproportionate representation within particular cohorts also suggests that selective admission criteria may significantly constrain the effectiveness of institutional outreach programs and other initiatives intended to increase diversity. The significance of family income and levels of educational attainment also predicts that there will be a continuing challenge to those selective institutions that wish to significantly extend admissions beyond the wealthy suburbs and professional communities of their respective states. The geographic locations that sent students to UVA well beyond their predicted values were located in some of the state's most wealthy and well-educated neighbourhoods, areas in which positional goods and public benefits may already be in abundance. Given emerging data on patterns of K-12 attendance and completion (McArdle, 2003; Teranishi & Parker, 2003, Kozol, 2006), this sharp differentiation is likely to continue.

The patterns of access presented here call for new approaches to state and institutional admission policies, including a renewed commitment to better understanding and enhancing the utility of such equity enhancing admission strategies as outreach, affirmative action, and percent plans. A number of flagship institutions, including the University of Virginia are in the forefront of new initiatives to increase campus diversity. As one example, UVA has created Access UVA, a program offering full tuition and other financial support to eligible low-income applicants from throughout the state. Other public and private institutions have also adopted similar policies, such as the University of North Carolina-Chapel Hill, Princeton University, University of Pennsylvania, Stanford University and Massachusetts Institute of Technology. In addition, institutions are waiving application fees, sending unsolicited 'personalized application forms' to targeted students, and offering quick turn-around time for admission decisions. In fall 2006, Harvard and Princeton Universities announced that they would discontinue their early admissions policies, citing that these programs reproduce advantage. Finally, some have even began soliciting targeted students in their eleventh year of secondary education, a full-year prior to postsecondary enrolment; though the National Association for College Admission Counselling recently updated its rules barring colleges and universities from offering students admission prior to September 15 in their twelfth year of secondary school.

This research also points to the need for increased attention to the definitions of equity and proportionality in post-secondary attainment. We have chosen to compare the proportion of students from various demographic and geographic sub-populations who apply, are admitted, and enrol at a state flagship to the proportion of individuals from those sub-populations in the state's entire 17-21 year-old population. This is one of many models that need to be pursued in future research. We have chosen the 17-21 year-old population in light of the fact that 99 per cent of the students in the admissions cohort at the flagship under study are from that age group. However, the 17-21 year-old age cohort is an increasingly small percentage of all post-secondary attendees and future studies may benefit from using different age cohorts for comparison.

In the final analysis we suggest that flagship universities are key sites for the production of public and private benefits of great value to individuals and communities. We further suggest that they serve as useful sites for the production of a key public benefit, the development of leaders. As public institutions they have an historical commitment to distribute that benefit as widely and as equitably as possible. As Thomas Jefferson, the founder of the University of Virginia, wrote:

> By that part of our plan which prescribes the selection of the youths of genius from among the classes of the poor, we hope to avail the State of those talents which nature has sown as liberally among the poor as the rich, but which perish without use, if not sought for and cultivated. (Jefferson, 1787, p. 247).

On the basis of the analysis provided here, it would appear that there is much still to be done in order to fully achieve that vision.

REFERENCES

Advisory Committee on Student Financial Assistance (2001). *Access Denied: Restoring the nation's commitment to equal educational opportunity*. Washington, DC: Advisory Committee on Student Financial Assistance.

Advisory Committee on Student Financial Assistance (2002). *Empty Promises: The myth of college access in America*. Washington, DC: Advisory Committee on Student Financial Assistance.

American Fact Finder (1990). Washington, DC: US Census Bureau. Available at: http://factfinder. census. gov /servlet/BasicFactsServlet

American Fact Finder (2000). Washington, DC: US Census Bureau. Available at: http://factfinder. census.gov/servlet/BasicFactsServlet

Astin, A. (1990). Educational assessment and educational equity, *American Journal of Education*, 8, 458-478.

Astin, A. & Astin, H. (2000). *Leadership reconsidered: Engaging higher education in social change*. Battle Creek: W.K. Kellogg Foundation.

Astin, A. & Oseguera, L. (2004). The declining 'equity' of American Higher Education. *The Review of Higher Education*, 27, 321-342.

Bowen, H. (1978). The returns to investments in learning. *Occasional paper series*, 7. Centre for the Study of Higher Education. Charlottesville: University of Virginia.

Bowen, W. & Bok, D. (1998a). *The shape of the river*. Princeton: Princeton University Press.

Bowen, W. & Bok, D. (1998b). Get in and get ahead: here's why. *Trusteeship*, November, 10-14.

Bowen, W.G., Kurzweil M.A., & Tobin, E. (2005). *Equity and excellence in American higher education*. Charlottesville: University of Virginia Press.

Breland, H., Wilder, G., & Robertson, N. (1986). Demographics, standards, and equity. Washington, DC: American Association of Collegiate Registrars and Admissions Officers, American College Testing, Educational Testing Service, & National Association of College Admission Counsellors.

Burkhardt, J., & Zimmerman-Oster, K. (1999). How does the richest, most widely educated nation pepare leaders for its future? *Proteus*, 16(2), 9-12.

Casteen, J. (2001). *The 2000-2001 President's report*. Charlottesville: University of Virginia.

Chang, M., Witt, D., Jones, J., & Hakuta, K. (Eds.) (2003). *A Compelling interest: Examining the evidence on racial dynamics in colleges and universities*. Stanford: Stanford University Press.

Cohen, A. & Brawer, F. (1996). *The American community college*. San Francisco: Jossey-Bass.

Colarelli, S., Dean, R., & Konstans, C. (1987). *Selective recruiting: College characteristics and job outcomes*. Annual meeting of the American Psychological Association, August. New York, August.

Engell, J. & Dangerfield, A. (2005). *Saving higher education in the age of money*. Charlottesville, VA: University of Virginia Press.

Geiger, R. L. (2004). *Knowledge & money: Research universities and the paradox of the marketplace*. Stanford: Stanford University Press.

Greenleaf, N. (1990). Overcoming barriers to educational mobility in nursing. *Nursing Forum*, 25(13), 23-29.

Giroux, H. A. & S. S. Giroux (2004). *Take back higher education: Race, youth and the crisis of democracy in the post-civil rights era*. New York: Palgrave Macmillan.

Hearn, J. (1984). The relative roles of academic, ascribed, and socioeconomic characteristics in college destinations. *Sociology of Education*, 57, 22-30.

Hebel, S. (2003). 'Percent plans' don't add up, *The Chronicle of Higher Education*, 49(28), A26.

Horn, C. & Flores, S. (2003). Percent plans in college admission: A comprehensive analysis of three state's experiences. Report from The Civil Rights Project at Harvard University. Cambridge, MA.: Harvard University. Accessed at: http://www.civilrightsproject.harvard.edu/research

Hurtado, S. & Wathington Cade, H. (2001). Time for retreat or renewal? Perspectives on the effects of Hopwood on campus. In D. Heller (Ed.) *The states and public higher education policy: affordability, access, and accountability*. Baltimore: Johns Hopkins University Press.

Hurtado, S., Milem, J., Clayton-Pedersen, A., & Allen, W. (1999). *Enacting diverse learning environments*. Report 26-8. Washington, DC: ASHE-ERIC Higher Education Report.

Jefferson, T. (1787). *Notes of the state of Virginia*. London: John Stockdale.

Jefferson, T. (1818). *Report of the commissioners appointed to fix the scite [sic] of the University of Virginia*. Richmond: John Warrock.

Joint Statement of Constitutional Law Scholars (2003*). Re-affirming diversity: a legal analysis of the University of Michigan affirmative action cases.* Cambridge, MA: The Civil Rights Project at Harvard University.

Kane, T. (1999). *The price of admissions*. Washington, DC: Brookings Institution Press.

Karen, D. (1988). *Who applies where to college?* Annual meeting of the American Educational Research Association. New Orleans.

Keith, S., Bell, R., Swanson, A., & Williams, A. (1985). Effects of affirmative action in medical schools: A study of the class of 1975. *New England Journal of Medicine*, 313(24), 1519-1525.

Kerr, C. (2001). *The uses of the university. 5th edition*. Cambridge, MA: Harvard University Press.

Kirp (2003). *Shakespeare, Einstein, and the bottom line: The marketing of higher education*. Cambridge, MA: Harvard University Press.

Kirst, M.W. & A. Venezia (Eds.). (2004) *From high school to college: Improving opportunities for success in postsecondary education*. San Francisco: Jossey-Bass.

Komaromy, M., Grumbach, K., Drake, M., Vranizan, K., Lurie, N., Keane, D., & Bindman, A. (1997). The role of black and hispanic physicians in providing health care for underserved populations. *New England Journal of Medicine*, 334(20), 1305-1310.

Kozol, J. (2006). *The shame of the nation: The restoration of apartheid schooling in America*. New York: Crown Publishers.

Lemann, N. (1999). *The big test*. New York: Farrar, Straus and Giroux.

Lucas, C. (1994). *American higher education: A history*. New York: Saint Martin's Griffin.

Marginson, S. (1997). *Markets in education*. Sydney: Allen and Unwin.

Marin, P. & Lee, E. (2003). *Appearance and reality in the sunshine state: The talented 20 program in Florida*. Report from The Civil Rights Project at Harvard University, Cambridge, MA. Accessed at: http://www.civilrightsproject.harvard.edu/research

Martin, I., Karabel, J., & Jaquez, S. (2003). *Unequal opportunity: Student access to the University of California*. University of California Institute for Labor and Employment. The State of California Labor, 2003. Paper ch04. Accessed at: http://repositories.cdlib.org/ile/scl2003/ch04

McArdle, N. (2003). *Race, place, and opportunity: Racial change and segregation in Boston metropolitan area: 1990-2000*. Report from The Civil Rights Project at Harvard University, Cambridge, MA. Accessed at: http://www.civilrightsproject.harvard.edu/research

McDonough, P. (1997). *Choosing colleges: How social class and schools structure opportunity*. Albany: State University of New York Press.

McDonough, P., Antonio, A. & Horvat, E. (1996). *College choice as capital conversion and investment: A new model*. Association for the Study of Higher Education, 31 October-3 November, Memphis.

McDonough, P., McClafferty, K., & Fann, A. (2002) Rural college opportunity: issues and challenges. o American Educational Research Association, March. New Orleans,

McMahon, W. (1999). *Education and development: Measuring the social benefits*. Oxford: Oxford University Press.

Milem, J. & Hakuta, K. (2000). Special focus: The benefits of racial and ethnic diversity in higher education. In D. Wilds (Ed.) *Minorities in higher education 1999-2000: Seventeenth annual status report*, p. 39. Washington, DC: American Council on Education.

Miller, S., Rivell, J., & Walker, B. (1991). Weighted admissions sandards at public flagship universities. *The Journal of College Admissions*, 130, 15-19.

Moy, E. & Bartman, B. (1995). Physician race and care of minority and mdically indigent patients. *Journal of the American Medical Association*, 273(19), 1515-1520.

Nickens, H., Ready, T., & Petersdorf, R. (1994). Project 3000 by 2000: Racial and ethnicity diversity in US medical schools. *The New England Journal of Medicine*, 331(7), 472.

Olivas, M. A. (1999). Higher education admissions and the search for one important thing. *University of Arkansas at Little Rock Law Review*, 21, 993-1024.

Ordorika, I. (2003). *Power and politics in university governance: Organization and change at the Universidad Nacional Autonoma de Mexico*. New York: RoutledgeFalmer.

Pascarella, E. & Terenzini, P. (1991). *How college affects students*. San Francisco: Jossey-Bass.

Pitter, G. & LeMon, R. (1991). *Limited access: A conflict between quality and opportunity*. Tallahassee: State University System of Florida.

Pusser, B. (2001) The contemporary politics of access policy: California after Proposition 209. In D. Heller (Ed.) *The states and public higher education policy: Affordability, access, and accountability*. Baltimore: Johns Hopkins University Press.

Pusser, B. (2004). *Burning down the house: Politics, governance, and affirmative action at the University of California*. Albany: State University of New York Press.

Pusser, B. (2006a). Higher education, markets and the preservation of the public good, in D.W. Breneman, B. Pusser, & S.E. Turner (Eds.) *Earnings From learning: The rise of for-profit Universities*. Albany: State University of New York Press.

Pusser, B. (2006b). Reconsidering higher education and the public good: The role of public spheres. In W.G. Tierney (Ed.), *Governance and the public good*. Albany: State University of New York Press.

Richardson, R. & Skinner, E. (1991). *Achieving quality and diversity: Universities in a multicultural society*. New York: American Council on Education/ Macmillan.

Rudolph, F. (1990). *The American college and university: A history, 2nd edition*. Athens: University of Georgia Press.

Slaughter, S. & Rhoades, G. (2004). *Academic capitalism and the new economy: Markets, state, and higher education*. Baltimore: Johns Hopkins Press.

Special report: African-American college graduation rates: Intolerably low, and not catching up with whites (2002). *The Journal of Blacks in Higher Education*, 37, Autumn, 89-102.

Spence, M. (1973). Job market signalling, *Quarterly Journal of Economics*, 87, 355-374.

Teranishi, R., Allen, W., & Solorzano, D. (2004). Opportunities at the crossroad: Racial inequality, school segregation and higher education in California. *Teachers College Record*, 106(111), 2224-2247.

Thomas, G. (1979). The influence of ascription, achievement, and educational expectations on black-white postsecondary enrolment. *The Sociological Quarterly*, 20, 209-222.

Tierney, M. (1984). Competitive structure of selected college student markets. *Colleges and University*, 59, 229-244.

Wightman, L. (2003). Standardized testing and equal access: A tutorial. In M. Chang, D. Witt, J. Jones & K. Hakuta (Eds.) *Compelling interest: Examining the evidence of racial dynamics in higher education*. Palo Alto: Stanford Press.

Wilds, D. (Ed.) (2000). *Minorities in higher education 1999-2000: Seventeenth annual status report*. Washington, DC: American Council on Education

Winston, G. J. (1999). Subsidies, hierarchies and peers: The awkward economics of higher education. *Journal of Economic Perspectives*, 13(1), 13-36.

Winter, D., McClelland, D., & Stewart, A. (1981). A new case for the liberal arts. San Francisco: Jossey-Bass.

Brian Pusser
Centre for the Study of Higher Education, Curry School of Education
University of Virginia, Charlottesville, United States of America.

J. Kirsten Turner
College of Arts and Sciences
University of Kentucky, Lexington, United States of America